The Business of Mining

The Business of Mining
Series Editors:
Odwyn Ifan Jones
Former Principal of the Western Australian School of Mines, Kalgoorlie,
Australia and Former Dean of Mining and Mineral Technology,
Curtin University, Perth, Australia

A.J.S. Spearing & Eric Lilford
Western Australian School of Mines, Curtin University, Perth, Australia

ISSN print: 2640–463X
ISSN online: 2640–4648
Volume 3

The Business of Mining
Mineral Deposits, Exploration and Ore-Reserve Estimation

Volume 3

Odwyn Jones
Emeritus Professor, Curtin University, Perth, Australia

Mehrooz Aspandiar
*School of Earth and Planetary Sciences, Curtin University,
Perth, Australia*

Allison Dugdale
*School of Earth and Planetary Sciences
Curtin University, Perth, Australia*

Neal Leggo
CSA Global, Perth, Australia

Ian Glacken
Optiro Pty Ltd, Perth, Australia

Bryan Smith
Bryan Smith Geosciences Pty, Perth, Australia

CRC Press
Taylor & Francis Group
Boca Raton London New York

CRC Press is an imprint of the
Taylor & Francis Group, an **informa** business

A BALKEMA BOOK

CRC Press/Balkema is an imprint of the Taylor & Francis Group, an informa business

© 2020, Curtin University

Typeset by Apex CoVantage, LLC

Exclusive Licence to publish granted by Curtin University to

CRC Press/Balkema

Library of Congress Cataloging-in-Publication Data
Applied for
Published by: CRC Press/Balkema
Schipholweg 107c, 2316 XC Leiden, The Netherlands

First issued in paperback 2023

ISBN: 978-1-03-257067-9 (pbk)
ISBN: 978-0-367-14894-2 (hbk)
ISBN: 978-0-429-05754-0 (ebk)

DOI: https://doi.org/10.1201/9780429057540

Publisher's Note
The publisher has gone to great lengths to ensure the quality of this reprint but points out that some imperfections in the original copies may be apparent.

Contents

Foreword by the Vice-Chancellor of Curtin University

The WA School of Mines, established in 1902, has been a core part of Curtin University since 1969 when it came on board to deliver mining education programs as part of what was then the Western Australian Institute of Technology.

As one of the first mining schools in Australia, it has adapted and expanded over more than a century to better support the education and research needs of the broader resources industry. In addition to retaining its mining roots, the School now also incorporates chemical engineering, mineral and energy economics and petroleum engineering.

Today, it holds a reputation as one of the best mining education and research centres nationally and internationally. The School was the top research performer in the 2017 QS international rankings, scoring 95 for citations per paper and 100 for H-index citations. And in 2018, Curtin University was ranked second in the world for Mineral and Mining Engineering in the QS World University Rankings By Subject.

And so, we are delighted to be bringing you this Focus Series, *The Business of Mining*, that captures the great wealth of knowledge that the WA School of Mines has to offer – in mine valuation and risk, ore bodies and mineral exploration, accounting and mineral marketing.

The series' editor and original author Emeritus Professor Odwyn Jones – who himself led the WA School of Mines for 15 years – has done an outstanding job in harnessing this expertise and making it available to students and professionals in the global mining sector.

Curtin University's mining and minerals graduates are recognised and sought after around the world. Many go on to become high-achieving, dynamic leaders in their field.

Curtin's postgraduate program in mining and minerals is a rich and rewarding course of study: block teaching, online teaching and hybrid models of teaching have all been implemented to stay current, offer students flexible modes of study and ensure that field trips and on-site study are integral.

I am, therefore, especially delighted to see *The Business of Mining* align with Curtin's MOOC of the same name, and for royalties from the series to go towards supporting educational programs at the WA School of Mines and the mining leaders of tomorrow.

Professor Deborah Terry AO
Perth, Western Australia

Foreword by the editor

This series of Focus Books originated in a text written some 20 years ago. The intention was to publish it as a textbook with the royalties from sales being directed in their entirety to Curtin University's WA School of Mines.

Having been left to languish, the original text was finally submitted some three years ago to Professor Sam Spearing, following his recent appointment as Director of the WA School of Mines, for consideration and comment. He was of the view the material was worthy of being updated for publication and an approach was made to CRC Press/Balkema, Taylor and Francis Group, who agreed to publish the updated text.

Subsequently, a contract with the publisher was finalised, on the understanding that material would be published as a series of Focus Books, and that all royalties from sales be directed to Curtin University to support undergraduate or postgraduate students wishing to study full time for one year or more at the WA School of Mines' campus in Kalgoorlie.

The complete set of three Focus Books will provide readers with a holistic all-embracing appraisal of the analytical tools available for assessing the economic viability of prospective mines. The books were written primarily for undergraduate applied geologists, mining engineers and extractive metallurgists and those pursuing course-based postgraduate programs in mineral economics.

However, the complete series will also be an extremely useful reference text for practicing mining professionals as well as for consultant geologists, mining engineers or primary metallurgists.

Each volume has a discrete focus. The first volume presents an overview of the mining business, followed by an analysis of project variables and risk, an overall coverage of the royalty agreements, pricing and contract systems followed by a final chapter on accounting standards and practises for the minerals industry.

Volume two discusses, in some depth, alternative means of assessing the economic viability of mining projects based on the best estimate of the recoverable mineral and/or fossil fuel reserves.

The third volume commences with "Our Earth, its Minerals and Ore Bodies", followed by a review of mineral exploration and sampling of mineral deposits. It continues with detailed sections covering the reporting of mineral resources and reserves in Australia, and concludes with the basic principles and application of the various methods of estimating the in-situ mineral resources and ore reserves.

These Focus Books should also be particularly useful for students enrolled in Curtin University's MOOC 'The Business of Mining'. This course, designed in collaboration with leading industry and educational experts, concentrates on the theory and practice of appraising the economic merit and valuation of mining ventures.

My sincere thanks to my daughter Sian Flynne, Business Manager for Curtin's School of Economics and Finance, for her consistent support, professional advice and typing of the entire text. Sincere thanks are also due to all co-authors for their generous assistance in updating, and where necessary, extending the text to meet current-day standards and demands. As co-editors, Dr Eric Lilford and Professor Sam Spearing have made invaluable contributions in reviewing all three volumes and providing overall editorial direction – in addition to being co-authors themselves.

The administrative procedures involved in dealing with all co-authors, university officers and the publisher were complete and made much easier by the friendly assistance and guidance of Ms Valerie Raubenheimer, Vice-President Corporate Affairs, and the assistance of Dr Andrea Lewis, publications consultant on behalf of the University.

Finally, a special word of thanks to Curtin University's Vice-Chancellor, Professor Deborah Terry AO, for supporting the venture and to my colleague Professor Sam Spearing for his motivation, energy and friendship in making sure this project was initiated and completed on time. Lastly, but by no means least, sincere thanks to my long-suffering wife for her consistent support and wise counsel.

Emeritus Professor Odwyn Jones AO
Perth, Western Australia

The Business of Mining: author bios

Mehrooz Aspandiar

Dr Mehrooz Aspandiar graduated from the University of Pune in 1987 with a BSc in Geology and then completed his MSc and PhD in Geosciences at the Australian National University.

He joined Curtin University as a lecturer on a joint position with the Cooperative Research Centre for Landscape Environments and Mineral Exploration (CRC LEME) in 2000 and became a full-time teaching staff member in applied geology in the WA School of Mines at Curtin University in 2008.

Dr Aspandiar was involved with the education and technology transfer programs of CRC LEME and coordinates and teaches into several University programs related to mineral exploration and mining geology. His research deals with mineral exploration in general and the application of regolith to mineral exploration in particular. He was a core member of the AMIRA Project Predictive Geochemistry in areas of transported overburden from 2006 to 2008.

As part of CRC LEME, Dr Aspandiar contributed to teaching short courses on regolith geology and mineral exploration for the Geoscience Honours Program of the Minerals Tertiary Education Council (MTEC) from 2004 to 2008. He subsequently coordinated and taught the Mining Geology short course at WA School of Mines Kalgoorlie for the Geoscience Honours Program for MTEC from 2009 until 2015.

He is currently the course coordinator for the undergraduate BSc (Applied Geology) course at Curtin University and serves as the Director of Teaching and Learning at the School of Earth and Planetary Sciences, Curtin University. He is a Member of Geological Society of Australia.

Allison Dugdale

Dr Allison Dugdale graduated from the University of Melbourne with a BSc (Hons) in 1986, and subsequently completed a PhD in Economic

Geology focused on the Bronzewing Gold Deposit at the University of Western Australia in 2002.

Dr Dugdale has 13 years' mineral industry experience throughout Australia, predominantly focused on mineral exploration with Newmont Holdings Pty Ltd at the Telfer Gold Mine and then with Western Mining Corporation in North Queensland and at the Leinster Operations in Western Australia. In addition, Allison has eight years' experience as a research academic at the University of Melbourne, where her research focus was on orogenic gold mineralisation and associated hydrothermal alteration.

She joined Curtin University in 2013 as a sessional lecturer and was promoted in 2015 to Senior Lecturer in economic geology, mineral exploration and professional practice. She is also a Member of Geological Society of Australia, Australian Institute of Geoscientists and the Society of Geology Applied to Mineral Deposits.

Ian Glacken

Ian Glacken is an experienced resource industry professional. Ian was trained as a Mining Geologist and Geostatistician in the UK, the USA and in Australia, obtaining a Bachelor's and two Master's degrees, along with a graduate diploma. He has professional memberships of the Australasian Institute of Mining and Metallurgy, the Australian Institute of Geoscientists and the Institute of Materials, Mining and Metallurgy (UK). Ian has worked for mining and exploration companies in gold, nickel, copper, uranium and base metals and has been a consultant since 1998. He is currently the Director of Geology at Optiro, a boutique consulting and advisory group based in Perth, Western Australia.

Odwyn Jones AO

Emeritus Professor Odwyn Jones commenced work as a mining trainee with the UK National Coal Board (NCB) in 1950 and was granted an Industry Sponsored Scholarship a few years later to read for a mining engineering degree at the University College of Wales, Cardiff, where he graduated with a BSc with first class honours. He then returned to the industry, obtaining his Colliery Manager's Certificate.

In 1957 he accepted the position of Lecturer in Mining Engineering at the Royal College of Science and Technology, Glasgow, which later became Strathclyde University. His part-time research, involving both laboratory work and field-testing at a local colliery, was sponsored by the NCB, and he graduated with a PhD from the University of Glasgow in the mid-1960s.

In 1970 he accepted the position of Principal Lecturer in Environmental Technology of Buildings at the Polytechnic of the South Bank,

London, before moving on to Bristol Polytechnic in 1973 as Head of Department of Construction and Environmental Health.

In 1976, Emeritus Professor Jones and his family moved to Western Australia where he took up the joint posts of Principal of the WA School of Mines, Kalgoorlie, and Dean of Mining and Mineral Technology at the WA Institute of Technology, which later became Curtin University.

In 1991 he transferred from Kalgoorlie to the University's main campus in Perth as Director University Development (International) and Director of the Brodie-Hall Research and Consultancy Centre.

Having retired from the University in 1995, he served as Visiting Professor at the Virginia Polytechnic Institute and State University, Blacksburg, for a term before assisting Perth's Central TAFE in developing its minerals and energy-related programs, where he stayed until 2001.

He is a long-standing Member of Engineers Australia, and Fellow of both the Australasian Institute of Mining and Metallurgy and the Institute of Materials, Minerals and Mining (UK).

He was Deputy Chairman (1981–98) and Chairman (1998–2013) of the Research Advisory Committee of the Minerals and Energy Research Institute of WA, a Statutory Authority operating under its own Act of Parliament.

His recent awards include an Officer of the Order of Australia in 2005 and the Australasian Institute of Mining and Metallurgy's Service Award for 2012–13. He was inducted into the Australian Prospectors and Miners Hall of Fame in September 2013.

Neal Leggo

Neal Leggo is a geologist with over 30 years' experience including management, mineral exploration, consulting, resource geology, underground operations and open pit mining. He has worked in a variety of Australian geological terrains and specialises in copper, gold, silver-lead-zinc and iron ore for which he has the experience required for code-compliant reporting. He also has experience with uranium, vanadium, manganese, tin, tungsten, nickel, lithium, niobium, gemstones, mineral sands and industrial minerals.

Mr Leggo provides a range of consulting services including code-compliant (JORC, NI43–101, VALMIN) reporting and valuation, technical studies, reviews and management of exploration projects. He offers extensive knowledge of available geological, geophysical, geochemical and exploration techniques and methodologies, combined with strong experience in resource estimation, feasibility study, development and mining of mineral deposits.

Mr Leggo has a BSc (Hons) in Geology from the University of Queensland and is a Member of the Australian Institute of Geologists and MSEG.

Bryan Smith

Bryan Smith graduated with a PhD from Melbourne University in 1966, was a Research Scientist with CSIRO Division of Soils until 1972 and an Exploration Geologist/Geochemist with AMAX based in Kalgoorlie for two years. He then spent 11 years with WMC also based in Kalgoorlie and three years with CSR Minerals based in Perth. He was General Manager Exploration for Aztec Mining in Perth for seven years, subsequently establishing his own consultancy: Bryan Smith Geosciences. This has been operating since 1994, mainly in Western Australia but also in the Northern Territory, New South Wales, South Australia, Queensland and Indonesia. Mr Smith is Member of GSA, AIG, AIMM and the Association of Exploration Geochemists. He was a Member of the Mining Industry Liaison Committee from 1993 to 2018, and a Member of MERIWA – later MRIWA – from 2005 to 2018. He was a council member of the Association of Mining and Exploration Companies from 1993 to 2018.

1 Our Earth, its minerals and ore bodies

Our Earth and its mineral and petroleum resources

Planet Earth

Our planet Earth was created 4.55 billion years ago from condensates in a cosmic gas cloud that was the resultant product of a supernova. All the material goods that are essential in our daily lives are provided for us by our Earth, which approximates a closed system with regards to matter. However, as our planet moves through space it is continuously bombarded by meteorites and several tons of cosmic dust each day. Hence, the Earth is gradually increasing in size by ongoing planetary accretion. Only energy is exchanged with space. Energy from space is sourced from the sun whilst at the same time energy is radiated from the Earth. Overall this exchange is balanced to enable the Earth to keep an overall stable temperature.

The Earth contains four key spheres the atmosphere, hydrosphere, biosphere and the geosphere (Figure 1.1). Mineral and petroleum resources that are vital to our everyday life are extracted from the geosphere. However, many of these resources developed as a result of the dynamic interactions between these spheres.

The composition of the Earth

From our perspective as humans living on the surface, the Earth appears to be static except for the occasional earthquake or volcanic eruption but in truth it is dynamic. The study of earthquakes (seismology) led to the understanding of the formation and propagation of elastic (seismic) waves within the earth. An earthquake produces three types of seismic waves: Primary (P-), Secondary (S-) and surface (Rayleigh) waves. The P- and S-waves are referred to as body waves. P-waves travel in a similar fashion to ordinary sound waves, which propagate in the same direction as that of the wave, and

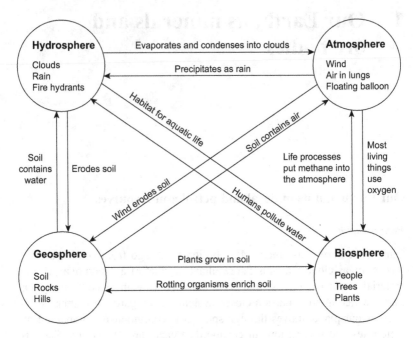

Figure 1.1 Illustration of the four Earth spheres: hydrosphere; atmosphere; biosphere and geosphere; and the dynamic interactions between the spheres.

are the fastest of the three waves (Figure 1.2). In contrast, S-waves move at right angles to the direction of wave propagation and travel at about half the speed of P-waves (Figure xx). P-waves can travel through both solid and liquid, whereas S-waves can only travel through solid material. The study of P- and S-waves generated by earthquakes has provided us with an understanding of the internal structure of the Earth.

Basically the Earth comprises three main concentric shells: crust, mantle and core (Figure 1.3). The crust is the rigid outer layer of the Earth and it comprises two distinct types: oceanic and continental. Oceanic crust is predominately composed of mafic igneous rocks (basalt and gabbro) and has an average density of 3.0 g/cm³ and an average thickness of 7 km. Continental crust has a variable composition but has an average density of 2.7 g/cm³ and a variable thickness, the average of which is 40 km but can be more than 70 km. The contact between the crust and the mantle is known as the Mohorovicic discontinuity (also known as the Moho) which was defined in 1909 by Croatian seismologist Andrija Mohorovicic. The Moho marks a change increase in the velocity of both P- and S-waves.

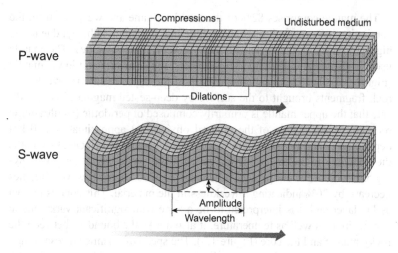

Figure 1.2 Seismic wave propagation of P- and S-waves.
Source: Smithsonian Ocean Portal.

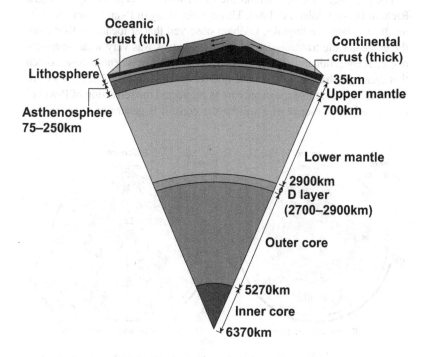

Figure 1.3 Illustration showing a simplified layered interior of the Earth.
Source: Ace Geography.

The mantle comprises 82% of the earth's volume and we know from the continuity of S-waves through the mantle that it is solid, although due to the high temperatures the rocks can flow at very slow velocities. The mantle itself is also subdivided into two sublayers: upper mantle and lower mantle (Figure 1.3). The chemical composition of the upper mantle is revealed by rock fragments brought to the surface by deep-seated magmas. These indicate that the upper mantle is primarily composed of peridotite (an ultramafic rock consisting mainly of the mineral olivine). At approximately 660 km depth both S- and P-waves show a significant increase in velocity; this marks the boundary to the lower mantle, which is the single largest layer occupying 52% volume of our planet. At depths of ~ 2700 km the S-wave velocities decrease by 30% indicating a weakness in the material. This zone is known as D" layer and it is interpreted to be a zone with significant variations in composition as well as temperature, it also marks the boundary between the rocky mantle and the core (Figure 1.3). The speed of seismic waves through the mantle, calculated from travel times, indicates an overall increase in rock density from 3.3 g/cm^3 in the upper mantle to 5.5g/cm^3 at the base.

The presence of a core within the Earth was first detected by geologist Richard Dixon Oldham in 1905. Through the study of P- and S-waves emitted from a large earthquake, Oldham observed that at locations 100° from the epicenter the transmission of P- and S-waves was very weak to absent respectively (Oldham, 1905). This was evidence that a central core existed that created a shadow zone for seismic waves (Figure 1.4). The absences of S-waves indicated that the outer core is liquid and the refraction of P-waves confirmed the presence of a solid inner core. It is generally accepted that

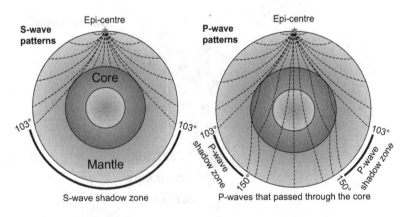

Figure 1.4 Seismic wave propagation through the Earth's mantle and core, showing development of S- and P-wave shadow zones.

Source: Physical Geology by Steven Earle is used under a CC BY 4.0 Licence, https://opentextbc.ca/geology/

both the inner and outer core are metallic, based largely on our knowledge of meteorites. The metal phase of meteorites contains approximately 94% iron and 6% nickel. It is, therefore, assumed that iron is the major component of the earth's core, with a small amount of nickel and another low atomic weight element such as sulphur, silicon, oxygen or carbon.

Plate tectonics

In the early part of the 19th century, meteorologist/geophysicist Albert Wegener introduced the hypothesis of continental drift (Wegener, 1912). He proposed that a single supercontinent (named Pangea) once existed but subsequently fragmented into continental blocks which drifted apart to their current position. Evidence in the form of fossils, matching coastlines, continuity of rock types and climate zones all indicated that Pangea had existed. However, there was no viable mechanism proposed to explain how the continents drifted apart after the fragmentation of Pangea. This was hotly debated until the 1960s when extensive oceanographic surveys recorded the presence of earthquakes localised beneath deep-ocean trenches and sampling of the oceanic crust found that there was no oceanic crust older than 180 million years. These discoveries lead to the formulation of the theory of plate tectonics.

The fundamental basis for the theory of plate tectonics is the recognition that the outer layers of the Earth are made up of seven major and numerous minor moving plates (Figure 1.5). Each plate comprises a crustal component

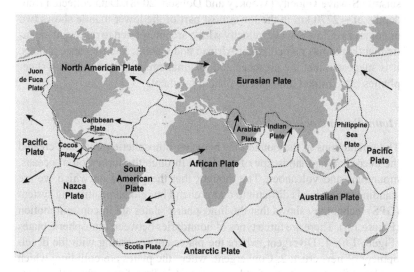

Figure 1.5 Map of the Earth showing the seven major tectonic plates and the current movement direction of the plates.

Source: http://sparkcharts.sparknotes.com/gensci/geology_earthsci/section4.php

and the upper part of the upper mantle, known as the lithospheric mantle. This combination forms a rigid outer zone called the lithosphere and hence the lithospheric plates. The thickness of the lithosphere in areas dominated by oceanic crust is approximately 100 km compared to 200 km in areas dominated by continental crust. The layer of the upper mantle beneath the lithospheric mantle is known as the asthenosphere, which is weaker (plastic) and denser compared to the lithosphere. Hence the lithospheric plates essentially float on top of the asthenosphere and these changes enable the plates to move independently from the asthenosphere.

The main driving force for the movement of the lithospheric plates comes from the Earth's internal heat energy, which is primarily driven by radioactive decay of elements such as uranium and residual heat from the formation of the planet 4.55 billion years ago. Within the mantle, heat is transferred by convection in which hot rocks rise upwards but as they cool they begin to sink leading to the formation of convection cells. In areas where the lithosphere is extended or thinned, the asthenosphere will be closer to the earth's surface which can focus the upwelling hot rocks which leads to the melting of the asthenosphere below the lithosphere boundary and the intrusion of hot primary mantle derived magma. The volume of the intruded magma forces the lithospheric plates to push apart and separate. The episodic nature of ocean basin opening and closing was first noted by John T Wilson in the early 1960s and is known as the Wilson Cycle (Wilson, 1963). Recent, seismic tomography maps of the Earth's interior show zones of fast and slow seismic S-wave velocity (Wookey and Dobson, 2008). Data collected from depths of ~ 2770 km in the lowermost part of the lower mantle shows two major areas of low S-wave velocity which are interpreted to represent gigantic mantle plumes named Great African and Central Pacific super plumes. It is suggested that the presence and magmatism associated of these super plumes may initiate lithospheric plate movement.

Mountains and oceans

The interactions at the boundaries between lithospheric plates has created the ever-changing landscape of our Earth through the formation of oceans, mountains and volcanoes. The thought that the ground on which we are standing is moving is difficult to comprehend but global positioning system (GPS) technology shows that the lithospheric plates are in constant motion (Figure 1.5). There are three types of boundaries between lithospheric plates (Figure 1.6): 1. Divergent, where the plates are separating with the development of new crust; 2. Convergent, where the plates are colliding, which results in the destruction of older crust; and 3. Transform, where the plates slide past one another along a major fault line. All of these boundaries are associated with increased seismic activity in the form of earthquakes.

THREE TYPES OF PLATE BOUNDARY

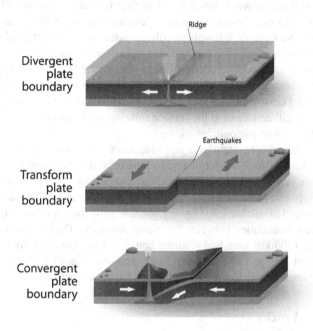

Figure 1.6 The different types of plate tectonic boundaries: a) Divergent, b) Transform and c) Convergent, with or without subduction.

Detailed oceanography of the deeper abyssal portions of the oceans has revealed a complex network of global mid-ocean ridge systems that extends for over 64,000 km. The ridge systems are divergent lithospheric plate boundaries, which are areas of intense volcanic activity where primary magma formed within the asthenosphere intrudes the thinner oceanic lithosphere pushing the plates apart and creating a new oceanic lithosphere (Figure 1.6a). The most obvious example of a divergent boundary is the formation of the Atlantic Ocean via the mid-Atlantic ridge. Approximately 200 million years ago all the continents of the Southern Hemisphere were joined together forming the southern portion (called Gondwanaland) of the supercontinent Pangea. Progressively over the next 200 million years North and South America separated from Africa and are still diverging today.

As the new oceanic lithosphere migrates away from the ridge systems the density of the lithosphere increases. When the density of lithosphere exceeds the density of the underlying asthenosphere, it will sink into the asthenosphere and this process is called subduction. The subducted slab of lithosphere will eventually be consumed by the mantle. However, volatile compounds (e.g. H_2O and CO_2 and S) trapped within the rocks will be released which can cause the mantle above the slab to melt and produce a magma that will rise upwards towards the surface forming a volcano. The subduction process may be spontaneous as described earlier or forced as is the case when the oceanic lithosphere collides with the continental lithosphere in convergence zones. In this situation, the oceanic lithosphere due to its higher density is forced beneath the lighter continental lithosphere (Figure 1.6c). The resultant magma above the descending slab pools beneath the thick continental crust which results in melting and formation of secondary magmas, which either ascend to the surface forming volcanos or crystallise at depth in the form of plutonic intrusions.

The mid-oceanic ridge system has a jagged appearance due to offsets caused by transform faults, which are dominated by horizontal displacement. Large-scale transform faults (> 1200 km in length) can also develop on the boundary between lithospheric plates where the plates converge at an angle that results in slippage instead of subduction (Figure 1.6b). Currently there are two major transform fault boundaries between the North American–Pacific (marked in part by the 1200 km long San Andreas Fault) and the Eurasian–African (marked in part by the 1500 km long North Anatolian Fault) lithospheric plates. These areas have a long history of major earthquake activity.

Minerals and rocks

A mineral by definition is a naturally occurring homogeneous solid, inorganically formed, with a definite chemical composition and an ordered atomic arrangement (Mason and Berry, 1968). Some naturally occurring minerals do not quite meet the requirements in terms of the full definition because they lack definite composition, a crystalline structure or both e.g. opal. Mineral-like materials are called mineraloid.

Minerals are divided into species based on the nature of the anionic group present. In total there are eight mineral species:

   I   Native elements
  II Sulphides (including sulphosalts)
 III Oxides and hydroxides
 IV Halides
  V Carbonates, nitrates, borates and iodates
 VI Sulphates, chromates, molybdates, tungstates
VII Phosphates, arsenates, vanadates
VIII Silicates

Of these different species, silicates are the most abundant, with 95% of the crust composed of silicate minerals.

Minerals are the building blocks of all rocks that make up the solid component of our Earth. In general, these rocks can be categorised into one of three fundamental types:

Igneous rocks   Solidified from the molten state

Sedimentary rocks   Formed by weathering of existing rocks and subsequent transportation, deposition and lithification

Metamorphic rocks   Rocks altered by the action of heat, pressure and fluids

Rocks and their environments

The oldest rock exposed on the surface of the Earth is a 4-billion-year-old igneous granitic intrusion in Canada. However, the oldest metamorphosed sedimentary rock, from Greenland, is 3.85 billion years old (Nutman et al., 1997), which tells us that there was an active hydrosphere present prior to the formation of this rock to enable weathering, transport and deposition of the sedimentary grains. It is hard to imagine that in the last 4 billion years entire mountain ranges and oceans have formed and subsequently being eroded, uplifted, deformed, metamorphosed, melted and recycled. In fact, all rocks and the minerals therein are related to one another via the rock cycle, which is influenced of two primary processes: internal processes (e.g. magmatism and metamorphism) driven by the Earth's geothermal energy and external processes (e.g. weathering, erosion and deposition) driven by the Sun's solar energy (Figure 1.7).

Igneous rocks are by far the most abundant rock type on our planet. Over 75% of continental crust and > 90% of oceanic crust comprises igneous rocks. These rocks are formed from the crystallisation of molten rocks called magma, which is derived from the melting of rocks in the mantle or the lower crust at temperatures > 800°C. The magma will either crystallise within the crust, forming intrusive (plutonic) igneous rocks, or will penetrate the surface of the Earth forming extrusive (volcanic) igneous rocks. Intrusive igneous rocks crystallise slowly with the development of large visible interlocking minerals, whereas extrusive igneous rocks crystallise quickly due to the extreme temperature differential between the magma and the atmosphere, which results in a limited growth of minerals and in some extreme circumstances development of volcanic glass (obsidian) with no minerals. The composition of igneous rocks provides us with clues as to the origin of the magma. Magmas that are derived from the mantle upon crystallisation form dark-coloured rocks (mafic or ultramafic) which have a high specific gravity and are dominated by iron- and magnesium-rich silicate

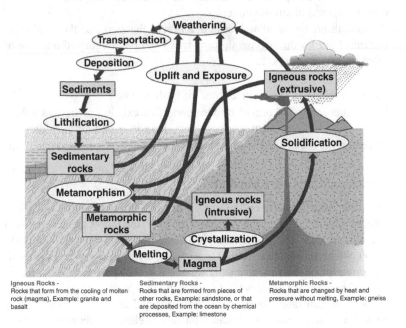

Igneous Rocks -
Rocks that form from the cooling of molten rock (magma), Example: granite and basalt

Sedimentary Rocks -
Rocks that are formed from pieces of other rocks, Example: sandstone, or that are deposited from the ocean by chemical processes, Example: limestone

Metamorphic Rocks -
Rocks that are changed by heat and pressure without melting, Example: gneiss

Figure 1.7 The rock cycle showing the various relationships between the various rock types in their formation and destruction.

Source: South Carolina Department of Natural Resources.

minerals. In contrast, magmas derived from the melting of continental crust produce rocks that are light-coloured (felsic), have low specific gravity and are dominated by silica-rich minerals. Magmas that are derived from the melting of the subducted oceanic crust in the mantle form crystalline rocks that are intermediate between the mafic/ultramafic and felsic extremes.

Sedimentary rocks typically form in areas of low to negative topographic relief with respect to sea level, where minerals, derived from pre-existing rocks, are subjected to weathering, transportation and eventual deposition and lithification. The pre-existing rocks may be igneous, metamorphic or sedimentary in origin. Sedimentary rocks are classified as clastic, chemical or organic. Clastic rocks are formed from the weathering of terrestrial rocks. The minerals and rock fragments from this process are transported by water, wind or ice over variable distances and finally deposited and lithified. Chemical sediments by contrast are developed as a result of precipitation from a fluid e.g. evaporites. Organic sediments are dominated by organic material e.g. shells, carbon.

Metamorphic rocks are developed within the crust. Metamorphism is a gradual process by which the minerals and texture of rocks, which may be igneous, sedimentary or metamorphic, change in response to increasing temperature and/or pressure with burial. Minerals that are stable at the earth's surface become unstable under specific temperature and pressure conditions and will undergo crystal lattice modifications to create a new mineral phase that is stable under the new conditions. This process will continue to occur with increasing depths within the crust. Commonly metamorphism occurs in response to regional scale deformation processes due to larger-scale lithospheric plate movements. Where the plates collide in convergent zones the crust is thickened in response to compressive forces much like squeezing an accordion. Metamorphic rocks in the deeper parts of the crust become like plasticine due to the elevated temperature and pressure, whereas weakly metamorphosed rocks in the mid to upper portions of the crust are brittle. So that in areas of compression (e.g. convergent lithospheric plate boundaries) the metamorphic rocks deep in the crust will fold and those in the mid to upper crust will fail, forming faults and thrusts.

Metallic mineral resources

Minerals have been extracted for use by humans for over 40,000 years, and it was the initial smelting of copper and tin from oxide ores and later iron that propelled civilization into the Bronze and Iron Ages. Metallic minerals in particular have a very special place in human history from their earliest uses in jewelry, coinage and weapons to current-day manufacturing industries producing products that are used by humans every day. The insatiable need for these minerals in the production of everyday items (e.g. mobile phones) has resulted in the elevation in price and competition to locate large enough resources for extraction.

Metals are typically hard, opaque and shiny, have good electrical and thermal conductivity and are generally malleable and ductile. Of the 118 elements of the periodic table, 91 are defined as metallic.

Some metals, such as gold, silver, platinum and copper, can occur as native elements, but most are formed as compounds in which metals are combined with one or more other nonmetallic elements and two of the most common being sulphur and oxygen. Metallic alloys also exist in which two or more metallic elements combine, e.g. electrum is a naturally occurring amalgam of silver and gold.

Concentrations of native metals, metallic compounds and alloys are unusual and anomalous compared to the average composition of the Earth's crust. These concentrations are commonly referred to as an ore deposit. Within any ore deposit there may be multiple ore bodies, which are bodies

Table 1.1 Clarke of Concentration for economically important metallic elements

Metallic element	Average concentration in upper crust (%)	Grade for typical economic ore (%)	Clarke of Concentration = enrichment factor
Aluminium (Al)	8.13	30	4
Iron (Fe)	5.00	20	4
Manganese (Mn)	0.10	35	350
Chromium (Cr)	0.01	30	3000
Copper (Cu)	0.006	0.25	40
Nickel (Ni)	0.0075	1.5	200
Zinc (Zn)	0.007	4	600
Tin (Sn)	0.0002	1	5000
Lead (Pb)	0.0013	4	3000
Uranium (U)	0.0002	0.1	500
Silver (Ag)	0.00001	0.05	5000
Gold (Au)	0.0000005	0.0005	1000

Source: After Ridley (2013).

of rock that contain elevated concentrations of metallic elements which may be viable for extraction. In order to be defined as an economic orebody, a metallic element needs to be enriched compared to its average upper crustal concentrations and the amount of enrichment is called Clarke of Concentration (Table 1.1). Common metallic elements such as aluminum (Al) have a low enrichment factor of 4, whereas tin (Sn) is very high at 5,000.

Classification of metallic ore deposits

Ore deposits can be classified on the basis of the predominant process that enabled the concentration of the metallic elements. These processes include magmatic, hydrothermal, sedimentary and regolith.

Magmatic ore deposits

Metallic elements within igneous rocks can be concentrated to economic concentrations through the processes of crystallisation (fractionation) in response to cooling of the magma. Typically the sequence of mineral crystallisation within magma depends on the composition of the magma itself, which results in the following magmatic ore types:

- Chromite (Cr) ores
- Nickel (Ni)-Copper (Cu) ores

- Platinum group element (PGE) ores
- Light rare earth element (LREE) ores

Metallic elements in magmatic ore deposits commonly occur due to the settling of early formed oxide minerals such as chromite, ilmenite and magnetite within the magma chamber e.g. stratiform chromite deposits in the Bushveld Complex, South Africa. Accumulation of immiscible sulphide melt droplets within mantle-derived magmas during crystallisation results in the scavenging of platinoid elements into the sulphide melts that form PGE reefs (e.g. Merensky Reef, South Africa; Figure 1.9). The incorporation of wallrock material during magma ascent or extrusion, particularly sulphur-bearing or silica-rich rocks, will result in the development of dense immiscible sulphide melts, which settle under gravity within the magma chamber or at the base of a volcanic flow. Deposits of this type include Ni-Cu deposits such as the gabbroic intrusion-hosted deposits (e.g. Sudbury deposit, Canada) and komatiite-hosted deposits (e.g. Kambalda deposits, Australia).

Pegmatite intrusions are the solidified liquid-product of extreme crystal-liquid fractionation of granitic source magma. The liquid portion of any fractionation of a magma becomes increasingly enriched in large ion lithophile elements (LILE, e.g. Li, Be), high field strength elements (HFSE, e.g. Sn, Nb, rare earth elements) and volatiles (e.g. B, F). Upon cooling these highly evolved magmas form distinctive coarse-grained pegmatites dominated by quartz, potassium, feldspar, muscovite ± tourmaline, topaz, tantalite, spodumene, cassiterite, apatite and lepidolite. Pegmatite intrusions are common, particularly in areas with significant granitic intrusive activity e.g. continental collision zones. However, some pegmatite intrusions can be significantly enriched in Li, Cs and Ta, referred to as LCT (e.g. Greenbushes, Australia) or in Nb, Y and F, referred to as NYF (e.g. Strange Lake, Canada).

Hydrothermal ore deposits

Hydrothermal activity involves the passage of aqueous solutions containing dissolved metallic elements through rocks. These fluids are derived either from depth (magmatism and/or metamorphism) or surface (meteoric, oceanic or connate–pore waters). Fluids derived from depth are sourced at pressures equal to lithostatic pressure but are overpressured with respect to fluids derived from the surface (e.g. groundwater) which have pressures close to hydrostatic. Fluid migration through rocks is enhanced by micro fracturing along grain boundaries in response to the pressure of the fluid. Fluids derived from depth tend to migrate upwards whereas those derived from the surface migrate laterally or convect. Products of hydrothermal fluid flow include hydrothermal alteration, veins and breccias. Hydrothermal alteration results in a change of chemical composition and mineralogy of the rock as a result of fluid interaction with existing minerals. The amount and nature of hydrothermal alteration depends on the permeability of the rock, existing mineralogy and the composition of the fluid. Veins develop as a result of fracturing of the rock due to either deformation or overpressured fluid. The opening of the fracture enables fluid to flow into the open space, which can trigger the crystallisation of certain minerals such as quartz and carbonate from solution. Breccias result when the pressure of the fluid exceeds lithostatic pressure, which can result in an explosive release of pressure and boiling of the fluid. Metallic elements dissolved within the fluid will precipitate either as a result of chemical reactions between the fluid and existing minerals or physical changes in the fluid in response to temperature or pressure fluctuations. Commonly ore deposits formed as a result of hydrothermal activity evolve over an extended time frame, sometimes in the order of > 100 million years. Hydrothermal ore deposits represent the most diverse style of mineralisation due to the variations in host rock lithology, metamorphic grade and degree of deformation and fluid composition.

The most common hydrothermal ore deposits are:

- Magmatic associated (Cu-Au ± Mo porphyry, VHMS, skarns etc.)
- Syn-orogenic (orogenic gold, IOCG, Carlin-style Au)
- Sedimentary (MVT, SEDEX, Red Bed Cu, Roll Front U)

Links between the various types of hydrothermal ore deposits have long been postulated (Figure 1.8) and it is common to have multiple deposits of different styles in proximity to one another, all of which are related to a single evolving hydrothermal system. All hydrothermal ore deposits exhibit metallic element zonation to some degree in response to progressive cooling of the fluid away from a conduit:

High Temperature (W, Sn, Mo) → (Cu, Au)
→ (Zn, Pb, Ag, Mn) Low Temperature

Sedimentary ore deposits

Mineral ore deposits formed as a result of sedimentary processes (known as hydrogene deposits) are dominated by some of the world's largest ore deposits as a direct result of precipitation from solutions on the Earth's surface. Iron, manganese and phosphorus deposits are the largest of this type formed from seawater precipitants. All of these deposits formed within shallow marine environments where the formation of the metallic ions is linked to interactions with an oxygenated atmosphere and/or deep oceanic upwelling (Figure 1.9).

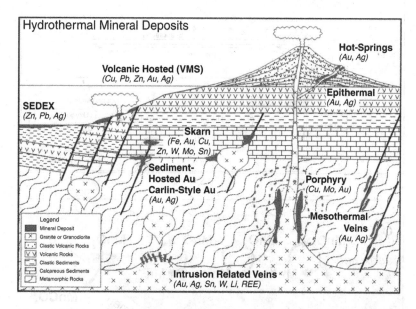

Figure 1.8 Schematic of the interrelationships between various hydrothermal ore deposits.

Other significant mineral deposits are accumulated during the clastic sedimentary process. Minerals collected in this way have the following characteristics: high density, high resistance to erosion and chemical stability. Minerals that exhibit these properties include the noble metals (Au, Ag, Pt), cassiterite (Sn), ilmenite, rutile, magnetite, wolframite (W) and zircon. Deposits of this type are called placer deposits. In active stream environments a sudden reduction in water flow velocity can result in the deposition of the heavy minerals (Figure xx). Other accumulations occur as beach deposits (Figure xx, whereby the wave motion of the sea on near-shore sands concentrate the heavier minerals on the landward site of the ripples called strands e.g. mineral sand deposits in Australia (Figure 1.10).

Regolith ore deposits

Rocks exposed to meteoric waters at or near the Earth's surface undergo chemical weathering of mineral constituents in the form of dissolution and leaching, hydrolysis and oxidation, and commonly these reactions are combined. The extent to which rocks are chemically weathered depends predominantly on the parent rock composition, climate and time. The resultant weathered rock

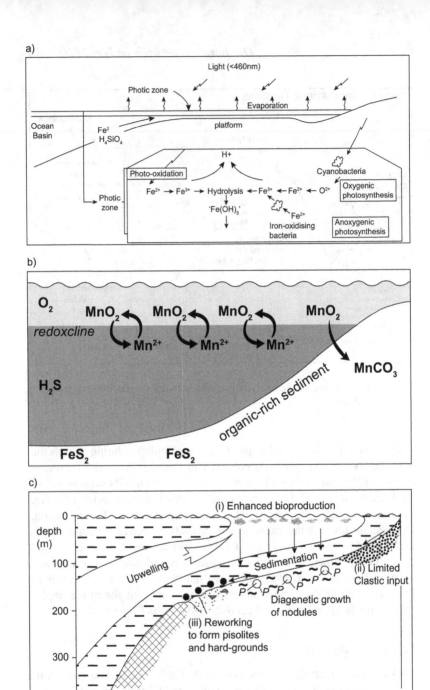

Figure 1.9 Potential mechanisms by which a) iron, b) manganese and c) phosphorous ore deposits may form in shallow marine environments.

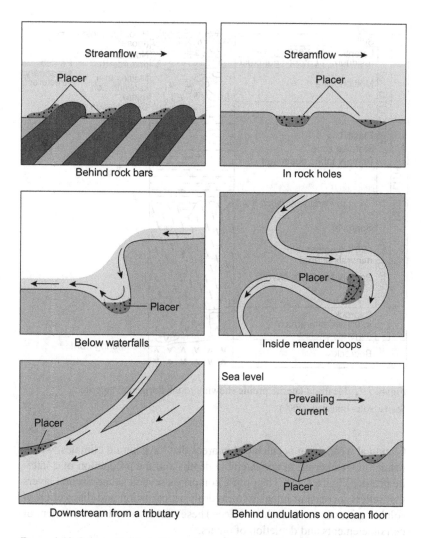

Figure 1.10 Schematic illustrating various methods of placer formation.

is termed *regolith*. Ore deposits formed as a result of chemical weathering are called supergene. Supergene ore deposits develop either as:

• Residual (in situ) ores
• Precipitation from shallow groundwaters
• Supergene enrichment

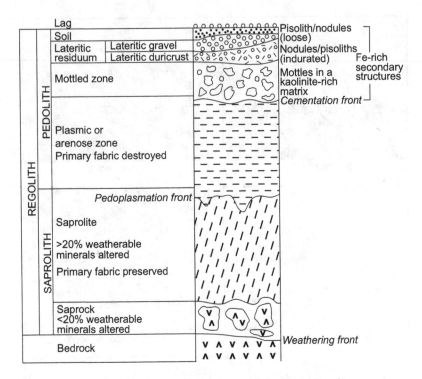

Figure 1.11 Idealised laterite profile showing rough horizontal layering.
Source: After Butt et al. (2000).

Chemical weathering is enhanced in areas that have humid climates, which results in weathering to considerable depths and the production of a laterite profile (Figure 1.11). The profile comprises several subhorizontal layers that reflect increasing intensity of hydrolysis, oxidation and dissolution of soluble minerals towards the surface. These processes cause enrichment of certain elements and depletion of others.

The enrichment of elements also results in the formation of residual ore deposits. The major types of these deposits include:

- Bauxite: dominated by aluminium hydroxide minerals developed from weathering of rocks that contain > 12% Al_2O_3 e.g. granite, arkosic sandstone e.g. Weipa deposits, Queensland.
- Laterite Nickel-Cobalt ores: dominated by Ni-Co rich clays (e.g. nontronite), serpentine and Ni-bearing Fe-hydroxide minerals (e.g. goethite) e.g. Bulong deposit, Western Australia.

Ore deposits formed as a result of the precipitation of minerals from shallow low temperature groundwaters include:

- Uranium deposits: hosted by calcrete or peat in currently active or buried alluvial channels in desert to semi-desert terrains e.g. Yeelirrie deposit, Western Australia.
- Magnesite deposits: nodules and concretions formed from weathering of ultramafic rocks e.g. Kunwarara deposit, Queensland.

Supergene enrichment deposits form from the redistribution, in response to weathering, of ore elements from an existing primary ore deposit. This process can result in a significant increase in grade of the ore element, which may be mined independently of the primary deposit. Secondary minerals commonly include oxides, carbonates, sulphates and native metals e.g. gold host the redistributed elements. Common supergene enrichment deposits include deposits for gold and copper.

Non-mineral resources (fossil fuels)

Coal has been used by mankind as a source of energy for centuries. The earliest evidence of coal mining occurred in China dated to 3490 BC. Coal was first documented as a useful component in metalworking by the Greek scientist Theophrastus (c. 371–287 BC). It wasn't until the Industrial Revolution during the 18th century that coal became an important commodity, particularly with the advent of smelting of iron ore using coke coal and the use of coal in steam engines. After the Second World War petroleum, natural gas and nuclear energy began to subside coal.

Coal

Coal is a combustible clastic sedimentary rock that contains > 50% by weight carbonaceous material, derived from plant remains, and moisture. The depositional environments where these sediments initially accumulate are typically marshes and swamps, where organic material can accumulate in stagnant anoxic water.

Coalification is a diagenetic process of burial ± tectonic activity that transforms moist, partially decomposed vegetation in response to progressively increasing temperature and pressure over time into a solid carbon-rich rock with minimal moisture. There are four main types of coal generated by coalification:

- *Peat*: Decomposed, spongy plant debris with high moisture content.
- *Brown Coal or Lignite*: Lignites contain recognisable woody material (lignin) but brown coals do not. Both have high moisture content and 40 to 55% volatile constituents.

•   *Bituminous Coal*: These coals burn with a smoky flame and contain 30 to 45% volatile content and are used for steam raising. Coking coals with 20 to 30% volatile content are used as smokeless fuels or to produce coke for the iron and steel industries.

•   *Anthracite Coal*: The transition from bituminous coal to anthracite is worked by loss of coking properties and decrease in volatile content. These coals contain up to 95% carbon and less than 5% volatile matter.

Coal is made up of macerals, which are organic material with botanical structures analogous to the crystallographic properties of minerals in inorganic rocks. Macerals can be categorised into three main groups:

•   *Vitrinite* is composed of woody plant remains: trunks, branches, stems, leaves and roots.

•   *Exinite* is composed of small organic particles such as plant spores, cuticle resins and algae. Common in low rank coals and oil shales.

•   *Inertinite* is oxidised organic material or fossilized charcoal.

Coal is commonly interbedded with layers of sandstone, siltstone and mudstone. The concentration of coal within one layer is referred to as a seam and many coal deposits will contain multiple seams.

Coal classification varies depending on country of origin, although it is based on a number of various properties including percentage of carbon, hydrogen and volatiles, the calorific (heating) value and the coking plus agglomeration properties (Figure 1.12).

Coal is analyzed for these properties using air-dried samples, which excludes surface moisture that is always present when coal is mined. Other analyses include the determination of ash content and the amount of nitrogen and sulphur.

Petroleum

Petroleum is a naturally occurring yellow-black mixture of liquid, gaseous and solid hydrocarbons hosted within predominantly sedimentary rocks. This material can be separated upon extraction by fractional distillation to produce natural gas, gasoline, kerosene, naphtha, fuel, lubricating oils, paraffin wax and asphalt. These products can then be further processed to produce a wide range of derivatives including plastic.

Phytoplankton converts atmospheric CO_2 into dissolved organic carbon, which is thought to be a major source of organic carbon deposited on the sea floor. Currently approximately 300 Mt of carbon is deposited in marine sediments each year. However, deposition rates were significantly higher during

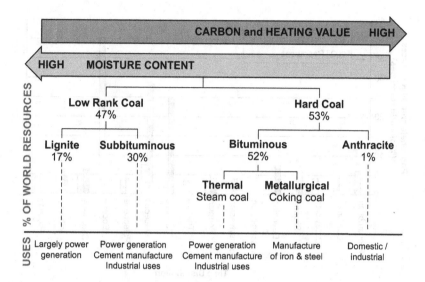

Figure 1.12 Coal classification scheme.

the Mesozoic period when atmospheric CO_2 levels were three times than the current level. This period coincides with the age of the primary source rocks for some of the major petroleum deposits (Figure 1.13). Commonly these source rocks are organic-rich shales that may also contain phosphate, sulphides and uranium. The conversion of organic matter into kerogen with the production of methane that occurs in response to physical, chemical and biological changes at low temperatures < 50°C is termed eogenesis. There are three types of kerogen produced:

- Type I Sapropelic – derived from algae in freshwater lakes
- Type II Mixed Planktonic – derived from marine algae and plankton
- Type III Humic – derived from terrestrial plant material

The dominant kerogen in the source rocks for the major petroleum resources is Type II kerogen, which is preserved in deep marine environments due to the anaerobic conditions on the sea floor and rapid burial.

The conversion of kerogen into petroleum and thermogenic gas occurs by a process known as maturation or catagenesis. With increasing burial (1000 to 3500 m) and subsequent heating (50–145°C), the large kerogen molecules break down to form smaller, lower molecular weight hydrocarbons (Figure 1.14). During this process the initial products oxygen and CO_2

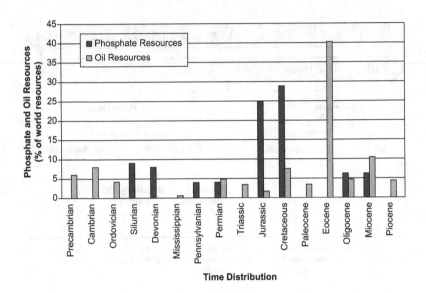

Figure 1.13 Age of petroleum and phosphate resources.
Source: Data compiled by the U.S. Geological Survey.

are progressively lost in response to dehydration. Heavy oils are the first petroleum products to form, followed by medium oils and at higher temperatures light oils and wet gas. Dry gas is only formed at temperatures $\geq 200°C$ and when the temperature exceeds 230°C all hydrocarbons are removed and only graphite remains. The depth and temperature range during which oils are produced is called the oil window. All three types of kerogen can undergo maturation to produce oil and gas.

The hydrocarbon products of the maturation process do not remain within the source rock and are flushed out by waters generated from the clay mineral reactions due to pressure gradients forcing fluid to flow upwards to downwards. This is called primary migration. Once out to the source rock, secondary migration results in the movement of hydrocarbons upwards through permeable rocks driven by buoyancy forces with denser connate waters (Figure 1.15).

The hydrocarbons continue to move upwards until they reach the surface, flow into a suitable reservoir or hit an impermeable barrier or trap (Figure 1.16). Rocks that are suitable for reservoirs typically have both high porosity and high permeability and the best reservoirs are sandstone and limestone. The traps that limit hydrocarbon migration are divided into three

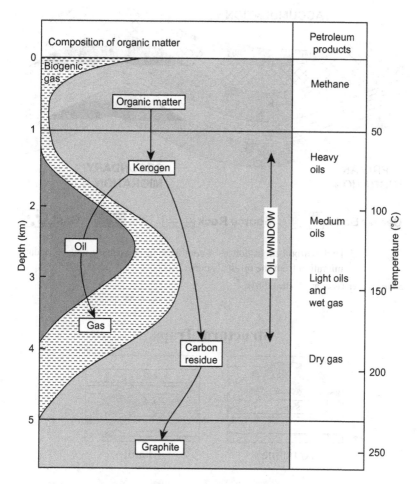

Figure 1.14 Summary of the progressive changes, with respect to increasing temperature and depth of burial, of organic matter to kerogen (eogenesis) and maturation of kerogen (catagenesis) to form hydrocarbons.

Source: Modified after Evans (1997).

main types: structural, stratigraphic and a combination of both structural and stratigraphic traps. Structural traps include anticlines or antiforms, faults and impermeable salt domes. Stratigraphic traps occur as a result of lateral and vertical variations in thickness, texture, composition and porosity of the reservoir rocks. Typically these changes result in the formation of wedge-shaped traps (Figure 1.16). The most common stratigraphic traps are related

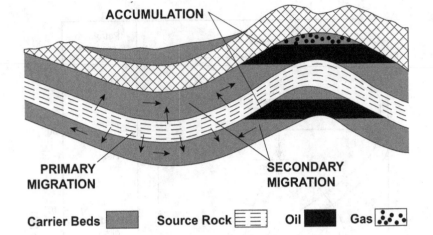

ACCUMULATION

PRIMARY MIGRATION

SECONDARY MIGRATION

Carrier Beds ▮ Source Rock ☰ Oil ▮ Gas ⦂⦂

Figure 1.15 Hydrocarbon migration as a combination of primary and secondary migration from the original source rock.

Source: After Tissot and Welte (1984).

Structural Traps

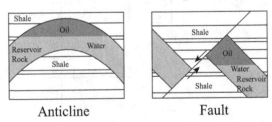

Anticline   Fault

Stratigraphic Traps

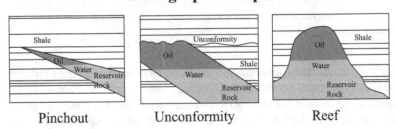

Pinchout   Unconformity   Reef

Figure 1.16 Structural and stratigraphic traps for hydrocarbon accumulation.

Table 1.2 Industrial minerals and their uses

Use	Mineral or rock
Abrasives	Quartz; Garnet; Corundum; Diamond; Sillimanite
Aggregates	Glacial till; Alluvial, beach and marine gravels and sand; Basalt
Structural clay products	Kaolinite, Illite, Smectite; Chlorite; Fluorite; Potash; Pyrophyllite; Sillimanite; Wollastonite; Andalusite; Feldspar
Cement and concrete	Limestone; Gypsum; Anhydrite; Pumice
Building stone	Limestone; Sandstone; Granite; Basalt
Glass	Quartz sand; Limestone; Soda Ash (Trona); Borates; Fluorite
Plaster	Gypsum
Paints	Rutile; Calcite; Feldspar
Electronics and Batteries	Graphite; Lithium; Mica; Talcl
Lubricants	Graphite
Pharmaceuticals	Magnesite; Soda Ash; Barite; Bentonite; Borax; Talc
Food	Phosphates; Salt; Sulphur

to unconformities where there is either a change in the structural attitude of the rocks or in the rock type either side of the unconformity (Figure 1.16).

Industrial minerals

Any rock, mineral or naturally occurring substance of economic value, excluding metallic ores, fossil fuels and gemstones, are defined as industrial minerals. The tonnage and total product value of industrial minerals far exceeds metallic ores. There is a vast variety of industrial minerals and their uses. The most common uses of industrial minerals are as abrasives, aggregates, structural clay products (bricks and tiles), cement and concrete, building stones, glass, paints, plaster, batteries, lubricants, pharmaceutical products and food (Table 1.2).

References

Butt, C.R.M., Lintern, M.J., and Anand, R.R. (2000). Evolution of regoliths and landscapes in deeply weathered terrain: Implications for geochemical exploration. *Ore Geology Reviews*, 16, 167–183.

Earle, S. (2015). *Physical Geology*, BC Campus, Victoria, B.C, https://opentextbc. ca/geology/

Evans, A.M. (1997). *An Introduction to Economic Geology and its Environmental Impact*, Blackwell Science.

Mason, B. and Berry, L.G. (1968). *Elements of Mineralogy*, W.H. Freeman & Co, San Francisco.

Nutman, A.P., Mojzsis, S.J., and Friend, C.R.L. (1997). Recognition of ≥ 3850 Ma water-lain sediments in West Greenland and their significance for the early Archean Earth. *Geochimica Cosmochinica Acta*, 61, 2475–2484.

Oldham, R.D. (1905). The rate of transmission of the Guatemala earthquake, April 19, 1902. *Proceedings of Royal Society of London*, 76, 102–111.

Pohl, W.L. (2011). *Economic Geology Principles and Practice*, Wiley-Blackwell.

Ridley, J. (2013). *Ore Deposit Geology*.

Tissot, B.P. and Welte, D.H. (1984). *Petroleum Formation and Occurrence* (2nd Edition), Springer, Berlin.

Ward, C.R. (Ed.) (1984). *Coal Geology and Coal Technology*, Blackwell Scientific Publications.

Wegener, A.L. (1912). Die Entstehung der Kontinente. *Geologische Rundschau*, 3, 276–292.

Wilson, J.T. (1963). Continental Drift. *Scientific American*, 208(4), 86–103.

Wookey, J. and Dobson, D.P. (2008). Between a rock and a hot place: The core-mantle boundary. *Philosophical Transactions of the Royal Society*, A 366, 4543–4557.

2 Mineral exploration

1 Introduction

Mineral exploration is the activity associated with searching, finding or detecting and defining an economically extractable ore. Exploration operations are generally made up of smaller areas such as districts and smaller prospects. A prospect is a defined area of ground that has a possibility of having an orebody (Marjoribanks, 2010) and, therefore, the main aim of mineral exploration activity is to narrow down the search area to better define targets. Prospecting uses a variety of methods from the traditional ones of searching for surface outcrops of mineralisation to the methods of conceptual geology, geochemistry and geophysics for exploration of tenements that have minimal surface outcrop.

Large-scale exploration requires planning, a significant budget and teamwork, and includes geologists using the latest geological models and interpretations, utilising the technological advances developed for geophysical data collection and processing as well as the latest geochemical analytical methods and interpretive techniques. The commodity and the area in which exploration proceeds depends on the:

- Access to funds for high-risk exploration programs.
- The availability of the social and environmental licenses to explore and eventually mine in the area selected for exploration.
- The political and legal risks associated with the country where exploration is conducted.
- Access to exploration leases and geological data, quality of the workforce, available infrastructure, weather and the metal commodity in demand driven by real or perceived future technological advances.

For example, resource companies are either set up to explore a single commodity or to change focus to explore for several commodities in current demand such as rare earth elements (REEs) that are required in high-end electronics, or for graphite as a source of graphene and battery anodes and lithium as a growing source in Li ion batteries for energy storage. Mineral exploration gets categorised into greenfields (or grassroots) exploration or brownfields exploration. Greenfields mineral exploration is conducted in regions that are geologically underexplored or not explored at all and so entail a higher risk and expenditure. Additional planning will be required and may entail access to areas with minimal or no basic infrastructure (Sillitoe, 2010). Greenfields exploration has the potential to provide the highest returns as often the best or largest ore deposits are the first ones to be found in a region. In contrast, brownfields exploration is conducted around existing or known ore deposits, and is considered lower risk as geological knowledge of the region and deposits are known (often referred to as 'mature terranes').

Exploration managers may refer to an exploration pyramid where the activity starts at the bottom with large areas, regions or camps that are identified based on geological factors. The base of the pyramid starts with conceptual models of ore formation then moves to a literature survey of previous work in the area, then to using the appropriate mix of detection methods of geophysics and geochemistry. The geological information may be sourced from the country's geological survey (or equivalent), ore deposit models and a mineral systems approach (Figure 2.1). Initial exploration within these broader areas may result in identification of geophysical and geochemical targets using concepts of anomalies but increasingly interpreting patterns in the data, which would be subsequently drill tested

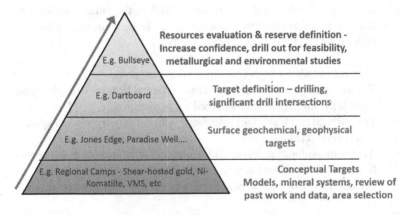

Figure 2.1 The exploration pyramid showing the gradual shift from large, reconnaissance targets to prospect scale targets to resourcing a deposit.

for mineralisation. If significant mineralisation or assay intersections are detected, then further detailed drilling would be conducted to check whether an economic resource can be identified. If the final resource is found to be economic, then a feasibility study would be initiated to define the technical and economic attractiveness of the deposit.

The quality and experience of the geological team is important. It is often said that conducting successful mineral exploration requires individual and team passion, perseverance and luck (Marjoribanks, 2010; Bennett et al., 2014).

2 The enrichment process and mineral systems – conceptual or prediction stage

Rocks and soils can contain a large range of elements, but most are at very low concentrations and so are not economic sources of commodities. The enrichment factor over the average crustal abundances for an element or ore to become economic ranges from a few multiples for bulk commodity elements such as aluminium (enrichment of approximately 2–3 times) and iron (3–5 times) to many orders of magnitude for precious metals like gold (1,000 times). In order to find ore deposits, we need to first understand the geological processes that cause metal enrichment in the Earth's materials and the geological locations where these processes have operated within or on the Earth's crust.

There are several geological processes operating throughout the Earth's history (mainly from 3.5 Ga to recent times) that have caused enrichment of metals in the upper crust. The main processes that are covered in Chapter 1 are due to differential crystallisation in magma, which can produce enrichment of elements including nickel, copper, platinum, palladium, lithium and rare earth elements (REEs). During hydrothermal processes, fluids present in the crust – such as seawater, deep groundwater and magmatic water – are heated and according to their chemical compositions, scavenge metals from spatial extensive crustal rocks and transfer them to another location, depositing the metals in a smaller or restricted location and thereby enriching the metals in that location. The hydrothermal processes are linked to metamorphic processes that may form gold, copper, lead and zinc deposits. The long-term chemical weathering of specific rocks in the uppermost part of the Earth's crust results in the enrichment of specific metals such as aluminium, iron, nickel, copper, gold, uranium and REEs. The resulting deposits are referred to as supergene deposits. Metal enrichment is also caused by sedimentary processes of deposition or precipitation. Such deposits are the placer deposits of heavy minerals such as rutile, ilmenite, zircon and gold, and chemical precipitates due to evaporation such as phosphates.

Most commercially exploited ore deposits have several geological requirements or processes that are important in the formation of ore deposits, thus defining ore body types and their models (Chapter 5, Cox and Singer, 1986; McQueen, 2005), and the processes have been expanded within the mineral system framework (Hageman et al., 2016). The main factors central to the formation of ore deposits – and therefore important to consider in a probabilistic manner – are the source of metals, medium to transport the metals, a throttle to focus the metals, a trap to deposit the metals, and a trigger or energy source to commence and mostly drive the entire process. The metal source is largely the partial leaching of metals from the mantle and crust. The medium to carry the metals from deeper locations (i.e. in the crust) is provided by fluids that are mainly magmatic, seawater, deep groundwater or metamorphic in origin, but may also be rainwater or shallow groundwater for supergene deposits. A transport or migration mechanism is generally mechanical or mass transfer through geological structures and other conduits like rock pores and rock contacts. Finally a throttle or focussing mechanism concentrates the metal-bearing fluids into narrow areas or trap locations, where the fluids deposit their metals mostly in the form of ore minerals. The final deposition of metals occurs due to changes in physiochemical conditions such as temperature, pressure, pH or redox (Eh) at the trap site. All the processes require a trigger, which in the mineral systems concept is a tectonic event to initiate and drive the entire system and process, but the trigger could be different for different types of deposits. For example, supergene deposits are not driven solely by tectonic processes but potentially by climate.

An example of the minerals system approach is one for orogenic gold type deposits (Figure 2.2). The trigger or energy for the system is acquired due to high heat flow and fluid flux from the mantle into the overlying crust. The source of fluids is either the mantle, magmas, metamorphic fluids or seawater, or meteoric water (hydrosphere waters). These fluids react with rocks along their path, scavenge metals and carry these metals upwards through conduits, which are crustal scale faults, structural domes or stratigraphic aquifers. The metal-bearing hydrothermal fluids encounter traps, which are physical barriers that cause the metals to be dropped or crystallised into minerals due to a change or drop in temperature of the fluid, the mixing of fluids or the reaction of the fluid with a rock. The reaction of the hydrothermal fluids with near-surface crustal rocks alters the rock and is known as hydrothermal alteration. This alteration concentrates the ore metals and minimises leakage of metals.

Many state and country geological surveys are increasingly employing the mineral systems concepts within their data acquisition and delivery

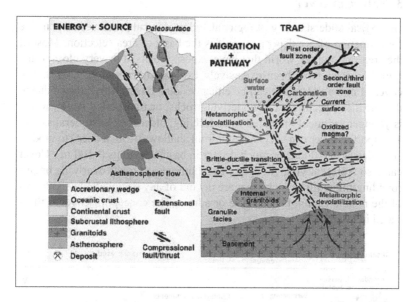

Figure 2.2 Minerals system summary diagrams for orogenic gold systems (from Hageman et al., 2016). The energy is provided by activity at convergent plate boundaries and the fluid source is provided by the metamorphism. The fluids are channelled along major structural discontinuities upwards and then into regional first-order shear zones (migration). The traps are individual second- and third-order structures, with deposition of gold facilitated by changes in fluid-rock reaction and phase separation.

program strategies and examples of these can be found at mineral systems framework of Geoscience Australia and the mineral systems atlas of the Department of Mines, Industry Regulation and Safety of Western Australia.

Brownfields exploration, as stated earlier, relies more on the knowledge of the local geology and known mineralisation to generate new targets or extend known mineralisation. For example:

- Structural orientation or plunges and structural controls in known mineralisation are interpreted to identify local targets from existing mines.
- Locally identified mineralogical and/or geochemical alteration criteria are used to direct lateral and/or deeper drilling.
- Stratigraphic marker beds or stratigraphic controls on mineralisation may be identified from existing mine geology and are used to generate extension targets.

3 Nature of cover

Geological understanding of regional and geological processes responsible for formation of ore deposits provides the basis of area selection. Most of the areas around the world that have exposed rocks at or very close to the Earth's surface have been extensively explored (Arndt et al., 2017). The cover of mineral deposits can be either residual cover, which forms due to the chemical weathering of underlying rocks, or transported cover, which is younger sediment unrelated to underlying, older rock or mineralisation. Transported cover can vary in depth from shallow (0–50 m deep) to moderately deep (50–100 m) or much deeper in basinal settings (>100 m). Cover can also represent younger age or even similar age rocks overlying or surrounding mineralised rocks that are devoid of any decipherable signatures of the mineralisation they overlie or adjoin. The concept of cover is illustrated in Figure 2.3.

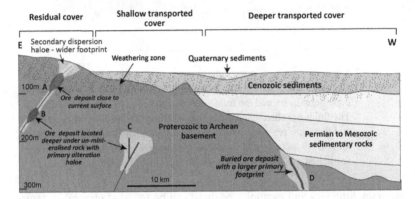

Figure 2.3 The concept of cover and concealed deposits in mineral exploration. Change in nature and depth of cover is shown from left (E) to right (W), where residual or in situ weathered cover is present as weathered zone profile, gradually moving to shallow transported cover (2–50 m) in the form of weathered to fresh mostly Cenozoic sediments to deeper (> 50 m) transported cover of fresh Proterozoic to Mesozoic sedimentary rocks. Magmatic and hydrothermal ore deposits have primary alteration haloes or footprints that are larger than the main ore, and secondary geochemical haloes or signatures near the surface and these are spatially more extensive than the narrower primary ore. Ore deposits can occur as extensions (B) of existing near surface deposits (A) that are targeted in brownfields scenario, or be concealed within un-mineralised basement rocks (C) or under deeper transported cover (D). The ore deposits illustrated do not represent any specific style of deposit and are shown for illustrative purposes.

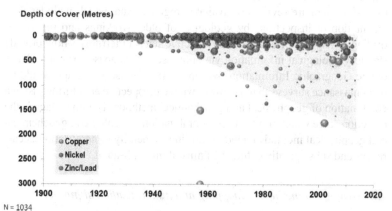

Depth of Cover (Metres)

N = 1034
Note: Size of bubble refers to "Moderate", "Major" and "Giant"-sized deposits.
    Excludes Nickel Laterite deposits

Figure 2.4 Depth of base metal deposits (Cu, Ni, Pb, Zn, except Ni-laterites) discovered since 1900.

Source: From Schodde (2014).

A survey of base and precious metal deposits discovered up to 2013 showed that most base metal deposits have been discovered at < 200 m (Figure 2.4), and large regions of the Earth's surface remain underexplored due to the presence of cover. Exploration in regions dominated by cover is considered to be the new search space (Schodde, 2014). To explore for concealed ore deposits obscured by a variety of cover types requires a sound understanding of geology but also the signatures of ore and techniques that detect those signatures. The signatures of ore have been traditionally detected via observations of surface geology and interpretation of rock, stream or soil geochemistry and geophysics, but given the variable depth and nature of cover, traditional and advanced geochemical and geophysical methods are increasingly being used. Irrespective of the method used, the interpretation of the data collected needs to be underpinned by geological principles.

4 Exploration techniques

Mineral exploration programs generally would follow a sequence or stages. The initial stage of exploration is likely to be the definition of the ore genetic processes to identify areas where there is a likelihood of formation of ore deposits – that is, selecting an area with a high probability of being a metal trap or an entire ore system. Much of the selection work is derived from extensive literature reviews of available regional data, which includes documentation of inspections by geologists of older workings and reports on drilling. Many countries and their respective states or territories now provide statutory geological information via online search databases or within basic online Geographic Information Systems (GIS). The second stage would be reconnaissance surveys that narrow down the prospective area further using a combination of geophysical and geochemical methods. The final stage is the detection stage which mostly uses a combination of geological, geochemical and geophysical methods including drilling to identify mineralisation occurrences and subsequently define the limits of mineralisation.

4.1   Geological methods – mapping and mineralisation signatures

The main objective of the detection stage is to identify surface and/or subsurface signatures of mineralisation that set an area apart from surrounding rocks or regolith. These signatures can be geological, geochemical or geophysical or a combination of these, but irrespective of the method used, a geological interpretation of the signatures is critical to the objective.

Traditional prospecting methods, where prospectors used visual clues to identify outcropping or near surface signature of ore such as altered rock and visible metal concentration occurrences, were responsible for over 50% of the discovery of base and precious metal deposits prior to the 1950s (Figure 2.5).

Geology is central to all exploration and besides providing conceptual models, geology is also used as a mapping tool that documents spatial location of rocks and geological features. Large-scale geological maps (1:250,000 scale) provide a base for the geological understanding of regions from where prospective ore system environments can be identified. Geological mapping can identify geological signatures of ore manifested in specific rock types of specific ages specific to their formative conditions. Mapping can also record mineralogical signatures indicative of occurrences of specific mineral types or assemblages. Large-scale maps provide the foundation for geological work while smaller-scale geological maps (1:10,000 or smaller) are used to document small-scale geological features. Most large-scale geological maps are provided by government surveys (for example, the Western Australian Department of Mines interactive geological map GeoView), although smaller-scale maps are usually prepared by company or consulting geologists. Although the value of geological mapping decreases in areas of cover, the principles used in surface mapping can be extended to sub-surface mapping using information acquired from drilling as well as geochemical and geophysical data sets.

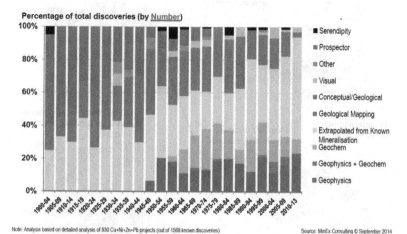

BASE METAL discoveries (>0.1 Mt Cu-eq) in the World: 1900-2013

Note: Analysis based on detailed analysis of 930 Cu+Ni+Zn+Pb projects (out of 1568 known discoveries)   Source: MinEx Consulting © September 2014

Figure 2.5 Primary search method at project scale to select the ground that led to discovery of Cu, Ni, Pb and Zn deposits.

Source: From Schodde (2014).

4.2 Geophysical methods

Geophysical techniques detect different physical properties of the Earth such as magnetism, gravity, electrical and radioactivity, and rely on the contrast between particular physical properties of the local or background rocks and those of target rocks to identify signatures related to ore (Dentith and Mudge, 2014). The geophysical response arising from the target geology can be higher or lower than the background surrounding geology and the deviation of the response is referred to as the geophysical anomaly (Figure 2.6).

Geophysical measurements represent a spatial series made in a spatial domain, where the geophysical variables measured (e.g. gravity, magnetics) are continuous. A series of measurements made over intervals of time is called a time domain, whereas measurements made at different frequencies are known as a frequency domain. Some geophysical methods use time and frequency domain measurements to provide information about the nature of sub-surface geology at the measurement location – which could be surface, airborne or down-hole (e.g. magnetics and gravity) – whereas other methods use it to provide information about geology at a distance from the surface (e.g. seismic).

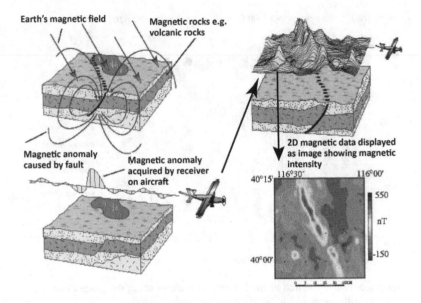

Figure 2.6 Concept of geophysical anomaly, data acquisition and presentation using magnetic as an example. A receiver on the aircraft measures the magnetic intensity from the Earth's magnetic field, and due to the presence of a fault or magnetic rocks, measures the high response or anomaly. A flight line will show up as a magnetic anomaly line but several flight lines will measure and display a two-dimensional magnetic intensity plot. The 2D data magnetic measurements plotted as a pixelated image shows the spatial variations in measured magnetic intensity.

Source: From USGS https://pubs.usgs.gov/of/1999/of99-514/about_method.html

4.2.1 Geophysical survey and processing data

The geophysical data interpretation starts with the geophysical survey, followed by data processing that results in the presentation of the data, culminating with the interpretation. In addition, geophysical modelling is also conducted to provide information on the source and geometry of anomalies.

The 'raw' geophysical parameter data collected is first subjected to a *data reduction step* where data is processed to remove known errors and minimise the noise. For example, in a magnetic survey, corrections are made to account for the influence of the Earth's magnetic field on the measurement. One of the data reduction methods is interpolating unevenly spaced 2D data points to a regularly spaced grid, with the process being referred to as gridding. Another correction is the suppression of noise. The noise is the undesirable variations in the geophysical parameter being measured and needs to be identified and reduced or eliminated, so the signal from the required geology is clearer. Methods to reduce the noise include repeated readings of the geophysical property, or the use of sophisticated signal processing algorithms such as Fourier analysis and filtering methods.

Geophysical data can also be enhanced to make the data clearer and useful. Some of the enhancements methods and the products that an exploration geologist will use are:

- Reduced to Pole images where the filters remove the asymmetry in the magnetic anomalies due the inclination of the Earth's magnetic field (most acute at equator).
- First vertical derivate (1VD) that calculates the vertical gradient of the magnetic field. It has the effect of sharpening the higher-frequency (short wavelength) magnetic field and is useful for enhancing the textural variation in the data including near-surface variations.
- Analytic signal (AS) image provides the envelope of the magnetic anomalies where the signal is unrelated to the direction of magnetisation.

4.2.2 Geophysical data presentation, products and modelling

Geophysical data are displayed in a variety of forms. The simplest data representation is the 1D profile where the variation in the physical parameter measured is displayed as a function of distance or time along the survey traverse line. These profiles show the anomaly shapes and possible presence of noise and errors in the data.

The representation of 2D geophysical data can be either as profiles, contours or images. The representations are in the form of line profiles of parallel survey lines or *stacked profiles*. A *contour plot* of the data depicts the spatial location of levels of constant amplitude across the area represented as contour lines. The contours are generally colour coded and the density of contour lines indicates the gradient. The most common form of 2D data representation is in the form of digital images where the intensity of the range of amplitude is represented by a greyscale or a range of colours (Figure 2.4). A variation of the image is shaded relief where the geophysical image is illuminated by light from any azimuth and elevation with amplitude variations being highlighted. The application of sun angle or sun shading helps to enhance geological features with different trend directions.

Geophysical data and its image representation are an indirect representation of the geology because the data are measured distant to the geology under consideration and the measurement is of a physical variable of the geology. Therefore, to assist in interpreting the geophysical response from the geology or target, models of the physical body are constructed. In geophysics, two types of models are used: forward models or inverse models. In forward modelling, the properties of likely geological section (depth, size, density – also known as model parameters) are used to make a geophysical model and compared to the actual geophysical measurements to achieve a reasonable match. In this type of modelling, the human interaction is crucial as the interpreter adjusts the model parameters to achieve a best fit to the actual data. This requires a sound understanding of the complexity of the model and the sub-surface geology by the interpreter. In inverse modelling, the process is inverted in that the actual recorded data or response is used to model the geology. That is, attempt is made to interpret the geology from the data. Inverse modelling is mostly automated and requires more assumptions about the model.

4.2.3 Individual geophysical techniques

The main geophysical methods used in mineral exploration are gravity, magnetics, electromagnetics and seismic and their basic principle and applications are summarised in Table 2.1.

Magnetic surveys are conducted both as airborne and ground surveys. In magnetic surveys the intensity of the Earth's magnetic field is measured or the amplitude of the magnetic field is measured at discrete points along a survey line on the surface. The main goals of magnetic surveys are to detect metallic ore bodies as these ore bodies show positive and intense magnetic anomalies and determine geological trends, patterns, extents and geometries of magnetic rock bodies, especially in covered terranes. The main minerals causing the magnetic signal are magnetite but also pyrrhotite and hematite. The normal product of a magnetic survey is in the form of Reduced to Pole (RTP) data and images of the magnetic intensity. Further derived products that add value to the magnetic data are available in the first vertical derivative (1VD) which is the vertical gradient and most common magnetic survey product after RTP, less so second vertical derivative (2VD) and analytical signal.

Gravity surveys provide a measure of the Earth's gravity field, which in turn is a subtle expression of variations in the rock density recordable with sensitive devices. The local minor differences in the mass of rock bodies produce an increase or decrease in the gravity field and these high and low mass variations as compared to immediate background mass show up as positive or negative anomalies in gravity data. Gravity is measured as milligal (mGal) and gravity meters measure very minor gravity differences. Gravity has generally been best measured via ground surveys, but developments in the past two decades have allowed better airborne gravity data to

Table 2.1 Principle and application of geophysical methods used in mineral exploration

Technique	Physical property	Applications
Magnetic	Measures intensity or strength of the Earth's magnetic field or magnetic susceptibility	Magnetite, pyrrhotite and hematite rich rock detection. Detection of metallic ores and structural trends.
Electrical resistivity	Electrical fields of rocks	Map sulphide ore bodies and define 3D geometry of host rocks.
Induced polarisation	Measures electrical surface polarisation of metallic minerals	Identification of disseminated sulphides (high response).
Electromagnetic	Electrical field generated and resulting secondary field measured	Detection of sulphide rocks where conductivity contrasts exist with host rocks. Also highlights structures.
Gravity (tensor gravity)	Measures Earth's gravity field, which is sensitive to variations in rock density with tensor being sensitive to small vertical gravity variations	Detection of denser sulphides and to map regional lithology and structure.
Radiometric	Measures the natural or induced radioactivity of the rocks/regolith	Detects radioactive minerals and primary alteration systems enriched in potassium.
Seismic	Measures reflection and/or refraction of elastic sound waves from rock	2D and 3D interpretation of discontinuities (structure) from crustal to local scales.

be collected. The airborne gravity surveys measure the gravity tensor and its popular version, the gravity gradiometer. The gravity method is used as a direct targeting tool in attempting to detect the higher masses of base metal sulphide deposits as compared to surrounding lower background mass.

Electrical methods employ the variations in the electrical conductivity of rocks and their constituent minerals. Electrical conductivity of rocks and minerals are measured in milliSiemens per metre (mS/m) and vary over orders of magnitude, although the variations of sulphides (mineralisation) can be similar to rocks and minerals that are non-mineralised, such as graphite and salt water. There are several electrical methods that take advantage of the specific electrical conductivity variations of rocks, with the main ones being induced polarisation and electromagnetics.

Induced polarisation (IP) measures the extent to which the Earth materials retain electrical charges or the chargeability of the material. In the IP survey, a current is applied to the ground and the voltage from the ground is measured. When the voltage decay is measured as a function of time, it is referred to as time domain mode, whereas when the current is applied to the ground at variable frequencies and subsequent voltages are measured to assess the different applied frequencies, it is known as frequency domain mode. The IP method provides good response from disseminated sulphide bodies and is, therefore, effective in identifying porphyry and stratiform deposits.

In the **electromagnetic (EM) induction** method, alternating current is passed through a transmitter coil, which produces a magnetic field known as the primary field, which then passes into the ground. The primary field in turn causes an alternating current to flow through the rocks, and depending on the rock types, the induced current in turn produces its own induced alternating magnetic field. The magnetic field produced is detected by the receiver coil at the surface. The method can measure the variations in the resistivity (or its inverse, conductivity) of the rocks from the surface to depths of several hundred meters. The conductivity measured is in units of Siemens/metre (S/m). As the method is an inductive one, no contact with the ground is required and surveys can be done on ground, in air (airborne EM – AEM) and on water, with the depth of investigation depending on the frequency and energy source of the transmitter and conductivity of the Earth. This method has been successful in exploring for massive sulphide ore bodies as the high conductivity (and low resistivity) of sulphides provides a clear response in contrast to surrounding less conductive rocks.

EM systems that place the receiver within the drill-hole to measure the conductive response are referred to as **down-hole electromagnetic (DHEM)** systems. The advantage of DHEM over surface or AEM is that the receiver being located closer to the electrically conductive target records a clearer response from the target, with interference from shallow conductors or large, weakly conductive bodies being minimised. This method also provides better resolution to detect closely spaced sulphide bodies that may appear as a single body from surface EM. Using a three-component DHEM probe, which measures two perpendicular cross-hole and axial components, it is possible to discretely identify the location and orientation of the sulphide conductor. This method together with forward modelling can be a successful tool in locating deeper sulphide conductive ores and is an invaluable tool in brownfields exploration for volcanogenic massive sulphide (VMS) style deposits.

Radiometric surveys deal with the natural radioactivity of the near surface. Although there are several radioactive elements present in the Earth's surface, much of the radioactivity present in regolith and rocks arises from the radioactive decay of only three main elements: potassium (^{40}K), uranium (^{238}U) and thorium (^{232}Th). Radioactivity is the process by which an unstable atom attains a stable state through the process of decay, or breakdown, of its nucleus. In achieving this decay, substantial amounts of energy are released in the form of α-particles, β-radiation and γ-radiation, of which only γ-rays travel farther than a metre in the air but are absorbed into the rock, i.e. only γ-radiation released from the upper 50 cm of the surface can be detected. Accordingly, an airborne radiometric survey of an area measures the spatial distribution of the three abundant radioactive elements (potassium K, uranium U, thorium Th) in the top 30–50 cm of the surface. Radiometric data are presented as ternary images where the multichannel (K, U, Th) data are assigned to three ternary colours: K to red, eTh to green and eU to blue. A combination of colours allows a semi-quantitative interpretation of the near-surface regolith and geology of an area. Enhanced images are constructed by using ratios of the elements such as eU/K and eU/eTh and are possibly the most useful in exploration.

Radiometric images only provide information on the upper centimetres of the surface. Therefore, this depth limitation needs to be considered during interpretation. Radiometric images allow the detection of potassic hydrothermal zones, which are rich in potassium-bearing minerals, biotite and illite. Radiometric surveys have been and are used to identify near-surface expression of U mineralisation, such as unconformity-hosted deposits and calcrete-hosted deposits.

For **seismic surveys** in mineral exploration (often referred to as 'hard rock' seismic), elastic waves generated by an acoustic blast propagated into the Earth from a source and its reflection and/or refraction are recorded by the detector or receiver in the form of geophones. The elastic properties of the rocks dictate the path of the seismic wave from the source to the detector and, depending on the rocks' elastic properties, waves will be deflected and will travel variably. The detector, therefore, records the different paths travelled by the waves. Analysing and identifying these different paths allows the inference of the nature of the sub-surface. Seismic surveys are of two types: first, seismic reflection surveys measure waves deflected at elastic discontinuities or geological strata (same angle reflection); second, seismic refraction surveys measure waves deflected but travelling parallel to elastic discontinuities or geological strata. Seismic reflection surveys at varying scales are increasingly used to provide information on major crustal boundaries that define large ore deposit terranes to unconformities, anticlines and faults on a camp to deposit scale.

Remote sensing methods use sensors mounted on satellites, airborne or proximal platforms to measure radiant energy reflected or emitted from the top micron of the Earth's surface across the electromagnetic spectrum from 350 nm–12,000 nm (0.35–12 μm). The sensors detect the energy to generate spectra and spectral features, some of which are diagnostic of the materials' composition (Cudahy, 2016). Sensors used range from satellite ones, such as Landsat TM and ASTER, to airborne hyperspectral ones, such as AVIRIS and HyMap, to proximal ones, such as handheld Portable Infrared Mineral Analyser (PIMA) and Fieldspec ASD, and to automated ones, such as HyLogger and Corescan's hyperspectral core imager Mark III. Minerals and mineral groups have diagnostic wavelength features, with the main ones being the Visible to Near-InfraRed (VNIR) which allows detection of oxides and hydroxides, Shortwave InfraRed (SWIR) that provides information diagnostic of hydroxyl-bearing minerals such as sheet silicates, carbonates and sulfates, and Thermal InfraRed (TIR), which potentially allows detection of several rock-forming minerals (such as feldspars, pyroxenes).

Satellite remote sensing is used in exploration as a regional or reconnaissance tool to assist in geological mapping and identifying detectable alterations via recognition of the hydroxyl-bearing alteration minerals suite, such as micas and chlorite (Sabins, 1999). However, satellite-based sensors have limited spatial and spectral resolution. Airborne hyperspectral sensors are mainly employed to identify alteration minerals, having better spatial and spectral resolution. However, most airborne and satellite sensors preclude detection of mineralisation signatures a few centimetres below surface, and in modern exploration, the value of spectral sensing is mostly around the use of proximal sensors such as ASD, HyLogger and Corescan that allow the identification and mapping of alteration minerals in drill samples (Cudahy, 2016).

4.2.4 Interpretation and targeting via geophysics

In cover-dominated regions, geophysical interpretations of the geology are used to define a target or reduce an area for further investigation.

In indirect targeting, geophysical survey data are used to interpret a suitable geological framework that is likely to identify the host mineral deposit. For example, the combination of magnetics and gravity is used to interpret lithological assemblages associated with specific ore deposits such as a broad scale alteration systems, or to interpret specific structural features that are necessary and suitable pathways and traps for ore metal-bearing fluids. Recent recommendations are to use geophysics, especially techniques that detect deeper deposits, to interpret ore systems – that is, to map out the pathway of fluids first and then concentrate on the trap (Witherly, 2014). For example, when shallow cover is widespread over a large region of the Yilgarn Craton, regional gravity and magnetics are good at distinguishing the lithological variations within the greenstones as well as highlighting major structures including shears, thereby aiding in narrowing down the search to smaller areas. Another example is where deeper seismic surveys allow interpretation of regional crustal scale structures and major terrane boundaries and domes that enable geologists to narrow down fluid migration regions.

In applying geophysics for direct targeting, a physical property of the mineralisation (be it conductivity, gravity or even geometry) is interpreted from various geophysical responses. For example, massive sulfide ore provides an EM signal due to its highly conductive nature, and such ore also provides positive gravity and density contrasts. In areas of known sulfide mineralisation, DHEM systems are used to define other sulfide-bearing ore. Magnetics are used to identify magnetite skarns whereas radiometrics are used to identify uranium-bearing mineralisation, especially shallow calcrete-hosted deposits.

An example of using a combination of direct and indirect targeting methods is the geophysical signature of copper-gold (Cu-Au) porphyry systems (Hoschke, 2011, Figure 2.7). Major porphyry deposits showed a strong magnetic signature due to the presence of magnetite as part of mineralisation (direct targeting), although the contrast between the host porphyry and non-magnetic distal alteration and surrounding rocks was important in highlighting the magnetic anomaly and its shape. Nature of alteration was important in the geophysical signature with potassic proximal alteration zone being magnetic, whereas the other distal alteration zones experienced magnetite destruction and were, therefore, displayed lower magnetic contrasts. Other aspects due to the general mineralisation and alteration related to the mineral characteristics of porphyry systems can also be detected – for example, strong to moderate chargeabilities (IP signatures) of a variety of sulfides such as pyrite and chalcopyrite within the mineralised core, and resistivity lows provided by clay minerals that result from hydrothermal alteration of the higher-resistive, unaltered surrounding volcanics.

Another example of brownfields type targeting is the use of newer 'hard rock' seismic for brownfields exploration in orogenic gold and nickel-sulfide ore systems. 3D seismic surveys are being used at prospect scales to first define indirect targets by defining structural traps such as anticlines, unconformities and major faults. Direct targeting using 3D seismic is also being employed on a prospect scale where reflector packages are being identified for drilling (Northern Star Resources ASX release, 20 February 2018).

4.3 Geochemical methods

Geochemical methods in mineral exploration rely on the principle that the chemical characteristics of rocks, regolith, minerals, groundwater, vegetation and gases are different within and around an ore body compared to elsewhere. The methods operate by detecting the geochemical differences

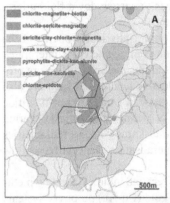
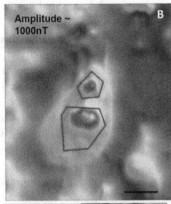
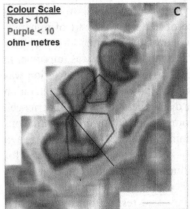
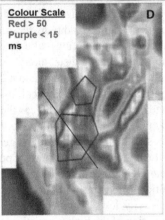

Figure 2.7 Elang porphyry Cu-Au deposit located in Sumbawa Island, Indonesia, showing geophysical response from three different methods to hydrothermal alteration associated with porphyry systems. A. Alteration map. B. RTP magnetics, showing magnetic anomaly in potassic zone due to magnetite alteration. C. The 200 m depth slice of resistivity that mirrors the alteration haloe. D. 200 m depth slice of chargeability (IP) that also shows the alteration of the porphyry but also the low due to dacite intrusion. The black polygons show the surface projection of magnetic bodies.

Source: From Hoschke (2011).

in the environment by carefully measuring the concentration of a single element or a suite of elements in the particular geochemical or sampling medium (e.g. rock, regolith type, vegetation, water, etc).

Geochemical methods are applied to detect chemical signatures arising from ores in two major environments: deep-seated and surficial. Deep-seated environments are where elements are concentrated or fractionated due to primary processes such as magma crystallisation and hydrothermal deposition. Surficial environments are where the Earth's surface processes of weathering, erosion and deposition disperse or concentrate elements. Linked to these two environments are the processes and types of geochemical dispersion. Primary dispersion is where the ore and pathfinder elements are dispersed due to magmatic or hydrothermal processes and their study is referred to as lithogeochemistry. Secondary dispersion is where the ore and related elements are dispersed due to secondary and surficial processes.

In exploration geochemistry, we tend to refer to target or ore elements and pathfinder elements. The target elements are those being explored for and that will be mined, such as gold, copper, nickel and uranium. Pathfinder elements are those that are not necessarily targets (such as arsenic, antimony, tungsten, selenium, tellurium, barium), but have important features that assist in finding the main source of ore. They might be associated in high or anomalous concentrations with the target mineralisation, or tend to disperse or spread wider than the target metals (be more mobile), thereby providing a larger ore signature useful for detection (Figure 2.8) and/or being easier to measure than the target. Some ore deposits' styles have specific target-pathfinder metal associations. In some ore environments, the target itself

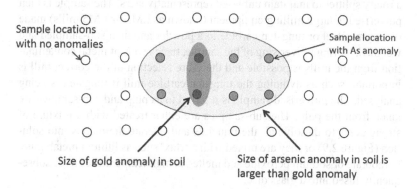

Figure 2.8 Concept of target and pathfinder elements, where arsenic as a pathfinder is firstly associated in higher concentrations with gold, but importantly is spatially spread over a larger area and detectable.

is considered the best pathfinder for itself (e.g. gold). In the example following, anomalous gold is restricted to only two samples and, therefore, spatially very narrow. In comparison, the arsenic anomaly from the ore is much wider and, therefore, easily detectable at a wider sample spacing. In any geochemical sampling program, it is good practice to use a multi-element suite that also includes the target element e.g. gold or copper along with a suite of pathfinder elements such as arsenic, antimony, molybdenum, tungsten, silver. The selection of pathfinders is either based on element associations with target and style of mineralisation (for greenfields exploration), or known element associations (for brownfields exploration).

4.3.1 Sample processing and analysis

All geochemical methods require the collection of samples (rock, regolith, vegetation, water and gas) and analysis of the sample to obtain a chemical concentration of a metal or metals that is reported in weight percentage, parts per million (ppm or g/t) or parts per billion (ppb μg/L). There are also other units used. The sample may be solid material such as rock or regolith. The sample, if large, will need to be reduced in mass and volume for analysis. Assays can be achieved in the field using portable analytical instruments such as portable X-ray Fluorescence (pXRF) instruments, but the detection limits are not as sensitive as those of laboratory instruments and representativeness of the sample is sometimes poor as compared to laboratory sample preparation.

In the laboratory analysis, the sample if coarse (>10 mm particle size) is first crushed using a Jaw crusher to achieve an optimum particle size before halving (splitting) the sample using either a rifle or Jones splitter or a rotary splitter to maintain unbiased representativeness. The sample is then pulverised in large milling equipment (known as LM2 or LM5 mills) made of chrome steel or tungsten carbide to a powder and then sub-sampled and weighed. During pulverising of the sample, trace levels of metal contamination from the mills is possible and therefore selection of the correct mill is important, such as avoiding the tungsten carbide mill if tungsten is being analysed. The pulverised sample is referred to as pulp and sub-samples are taken from the pulp. The sub-samples are either treated with a mixture of strong acids to dissolve all the minerals and liberate the metals into solution (Figure 2.9) or they are mixed with a 'flux' such as lithium metaborate/tetraborate or sodium peroxide and melted at high temperatures to be subsequently fused into a glass disc.

Partial and selective dissolution methods use weak acids or chemical reagents to selectively dissolve some minerals (Figure 2.9). Some of the leaches used are as simple as deionised water, weak-strength acids (such as 1 molar hydrochloric acid (HCl), weak sodium cyanide (NaCN) for

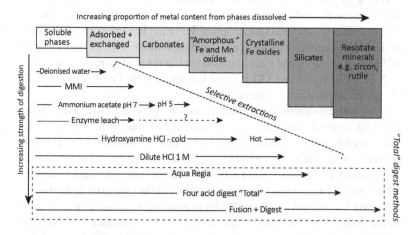

Figure 2.9 The different digestion methods commonly used to dissolve a rock or regolith sample. Note the link between the digest and phase(s) attacked by the digest.

complex fine-grained gold (referred to as Bulk Leach Extractable Gold, or BLEG), or sodium dithionite to preferentially dissolve iron oxide and hydroxides. Some digests are proprietary such as Mobile Metal Ions (MMI). Figure 2.9 shows commonly used selective digests and a full list of partial and total digest methods can be obtained from commercial laboratories. The concept behind employing partial and selective leaches is to increase the signal to noise ratio between anomaly and background, but also to map a fraction of an element that has transferred through transported cover and so will be loosely trapped or held in the regolith material matrix (Cameron et al., 2004).

A stronger sample dissolution technique is aqua regia (1 nitric acid: 3 HCl), a highly oxidising reagent, which is commonly used for gold analysis as it dissolves sulfides, native Au, tellurides, carbonates and most sulfates. Resistant minerals such as barite, chromite, ilmenite and rutile are not dissolved. For 'total dissolution,' a four-acid digest comprising hydrochloric acid (HCl), nitric acid (HNO_3), perchloric acid ($HClO_4$) and hydrofluoric acid (HF) is used. Although the four-acid digest dissolves most common minerals, some refractory minerals like barite, zircon, rutile, cassiterite, and tungsten minerals may not be dissolved. When a complete dissolution of the sample is required including resistate minerals, the fusion method of mixing the sub-sample with lithium metaborate flux and preparing a disc followed by dissolving the bead in nitric acid is followed.

Where digestions are used, the elements present in a liquid are analysed using analytical instruments such as Atomic Absorption Spectrometer (AAS), Induced Coupled Plasma Optical Emission Spectrometer (ICP-OES) or the highly sensitive Inductively Coupled Plasma – Mass Spectrometer (ICPMS). The major and trace elements within fused discs can be measured by X-Ray Fluorescence (XRF). Some laboratories use Laser Ablation ICP-MS on fused disc for trace metal analysis.

Analysis for precious metals such as gold are generally conducted via the fire assay method, where the sample pulp of between 10 to 50 g is mixed with soda ash, borax, lead oxide (litharge), flour and silica and fired at temperatures of approximately 1100°C in a furnace under reducing conditions. The reduced molten lead, due to its high density, filters through a mixture collecting gold with it and settles at the bottom of the container. This mixture is poured into mould and cooled. The mould that is glassy is then broken with a hammer and the lead button at the bottom is separated from the flux (glassy) and placed in a cupel. The cupel with the lead button is placed in a muffle furnace at 950°C to remove the lead. A small gold bead is left behind (Au) (often hardly any if minimal gold is present in the sample). The Au is either weighed or dissolved and determined by AAS or ICP-MS. A variant of the fire assay method for coarse or nuggety gold-bearing samples is the screened fire assay.

Besides using partial or selective digests, grain size separation followed by aqua regia or total dissolution is also followed. Recent developments in grain size separation workflow separate out the ultrafine or clay fraction of the regolith material followed by aqua regia digest and analysis and is showing promise in gold exploration through transported cover (Noble et al., 2018).

One important geochemical measurement parameter, especially in exploration, is the detection limit (DL) of the analytical procedure. Detection limit is the smallest net signal or derived lowest concentration of the particular element that can be distinguished from the background signal by the particular instrumental method. For example for gold, the analytical detection limit for the fire assay method with an AAS finish the detection limit is 10 ppb, whereas if measured by ICP-MS it is lower at 1 ppb (or even parts per trillion).

Technological advances in hardware and software have seen the advent of rapid analytical tools that use portable XRF instruments to measure concentrations from unprocessed samples such as core and drill cuttings. New methods using high energy X-rays (photon sources) to excite atomic particles of specific metals to then measure them from raw samples without any pulverising and dissolution are moving from the testing phase to the application phase (see Chrysos Corp).

4.3.2 *Quality control and quality assurance of assays*

Any assay acquired via analytical methods from the laboratory needs to be reliable and, therefore, a Quality Assurance and Quality Control (QAQC) system must be in place for the entire sampling process. Quality assurance relies on systematic actions necessary to provide adequate confidence in the data collection and estimation process.

For quality control, we need to measure precision, accuracy and bias. Accuracy is the degree to which an analysis or mean of a set of analyses approaches a 'true' concentration. Precision is the estimate of the reproducibility of the sampling and analytical system. Bias is a systematic error inherent in a method or caused by some artefact of the sampling system. Contamination is the introduction of any substance to a geological sample that is not in the original location of that sample. Materials of known metal content are known as standards and these are used to assess the accuracy and precision of assay results. Duplicate samples are usually inserted into the sample stream at the rate of 1 in every 20 to 1 in every 50 samples to assess the precision (i.e. repeatability) of assay results. Samples of nil content of a metal being assayed are called blanks and are used to assess laboratory contamination.

4.3.3 *Data analysis and presentation*

The main purpose of data analysis in exploration geochemistry is to identify anomalies so that geochemical targets can be generated. There are different definitions of anomalies, and a simple one is that anomaly is the deviation from the geochemical patterns that are normal for a given area of geochemical environment. Anomaly is an abnormally high or low concentration of an element or element combination, or an abnormal spatial distribution of an element or element combination in a particular sample type in a particular environment as measured by a particular analytical technique. Significant anomalies serve as a guide to ore. False anomalies arise due to high concentrations of pathfinder elements unrelated to ore. Anomalies are best viewed as signifying different populations within a particular geochemical data set and, therefore, it is necessary to have a significant number of samples from which to identify populations, otherwise the anomalies could simply signify random high data values.

Most exploration geochemical data analysis, irrespective of the method employed, should commence with first checking or validating the geochemical data. The first step is to check the quality of the data – whether it is fit for purpose. If there are issues with precision, accuracy or contamination of the geochemical data, go no further! Repeat the analysis. Also check for inconsistencies. Remove internal standards so they do not accidentally show up in the analysis. Compare the data ranges with the detection limits (DL) for each element. Often values at or below DL are changed to DL value or a consistent proportion of the DL (e.g. half DL). Also check the units of analysis (ppm, wt%, ppb).

The second step is assembling or merging the data. If the spatial location data (field data) and the geochemical data (acquired from laboratory) are from different sources, it is important to correctly merge the data. Furthermore, where available, geological and regolith attribute data (sample type, depth etc.) should also be merged. Record the analytical method (XRF, pXRF, ICP-MS) and digestion process used for the samples (partial, total, aqua regia, fusion, etc.).

The third step is to look closely at the data. This is best achieved via Exploratory Data Analysis (EDA). This type of data analysis assists in recognising anomalies and providing insights into geochemical and geological provinces and processes (Grunsky, 2010). The first step is to generate statistical summary tables (ranges, mean, media, standard deviation) and then to view the individual element data distribution using a number of visual aids such as histograms, probability plots, box plots and density plots. Probability plot is an effective means to visually identify populations and outliers within the data set. Often, univariate analysis plots are sufficient to achieve the aim of the geochemical survey to identify anomalous areas.

Together with EDA analysis, spatial presentation (geographic context) is necessary, and GIS is now standard software to visualise spatial distribution of geochemical data and aid in its interpretation by allowing merging and visualisation of other data sets. A simple method of visualising variations in the geochemical data is to provide a symbol for each location (several symbols if several attributes exist such as soil, basalt, granite etc.), and scale the symbol either linearly or exponentially to reflect the selected interval ranges of the data. For example, the 95th percentile of data is a simple way to visualise 'first pass' anomalies in the data set and can be achieved by sizing symbols for the locations that have element values above the selected percentile value. Another visual enhancement of the data is by providing colour and scale to the locations based on selected interval ranges. A final spatial product is the grid or contour image where the contour lines and designated colours define constant values. The contour image, however, interpolates data and may not be a true representation of selected element distribution. The most effective visual methods are to use EDA plots and spatial plots in concert, where the EDA plot allows the judicious method to select the anomalous population samples to be then plotted spatially. However, the traditional concepts of searching of anomalies in background populations needs to be approached with caution as background varies across scales from regional to prospect, and multivariate analysis needs to be increasingly used to interpret and tease out mineralisation indicators as opposed to local geology or anthropogenic-induced geochemical anomalies.

Multivariate data analyses, where two or more element variables are employed to calculate or visualise the data, are valuable to decipher relationships between elements and interpret geochemical processes from survey data. Scatter plots are the simplest multivariate visualisation tool and invaluable to view relationships between two or more variables. They are often used extensively in lithogeochemistry to classify rock types and identify hydrothermal alteration. Several multivariate methods are employed – from the simple but effective scatter plots (two or more variables) to correlations coefficients to more sophisticated Principal Component Analysis (PCA) and cluster analysis. Cluster analysis of the data set is employed when relationships in multi-element data are not easily decipherable from scatter plots.

4.3.4 Drainage or stream sediment

The oldest (in use since Roman times) and most cost-effective geochemical method in mineral exploration is stream sediments or drainage geochemistry, although its effectiveness is restricted to actively eroding landscapes or erosional regimes, and therefore not applicable to deposits concealed beneath cover. It is a technique successfully employed from regional reconnaissance to prospect scales.

The principle of the method relies on material in streams, especially sediments varying in sizes from clay to boulders, representing Earth material weathered and eroded (sediment) or derived in solution from its catchment. A metal source (a likely ore deposit) experiences chemical and physical weathering and due to hydromorphic and mechanical dispersion, metal from the source is dispersed into the surrounding soil or dispersed through waters draining the soil. Due to slope processes (gravity-based), erosion of the soil materials occurs downslope into the streams, thereby facilitating dispersion of the metals. The extent of dispersion of the metal into the streams is referred to as dispersion train, and the intensity of the dispersion decreases or decays downstream away from the source. Geochemical analysis of material at selected points along the length of streams will reflect the chemical composition of the rocks, regolith and ore from their catchment.

To conduct a drainage survey, a drainage map of the area of interest is prepared with drainage divides marked followed by the selection of stream sediment collection locations mainly along low-order streams before the confluence of two streams (Figure 2.10).

The grain size of sediments needs to be considered carefully as best results reflecting source are dependent on selection of grain sizes, often

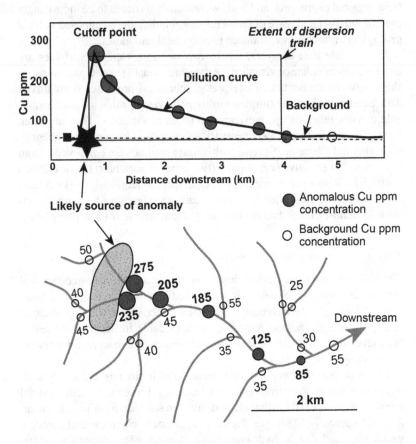

Figure 2.10 Concept of drainage surveys with example of anomaly dilution and identifying source region for an anomaly. The upper figure shows a section of Cu values decreasing away from the source due to dilution and the distance from source of detectable Cu is the dispersion train. The lower image shows a drainage map of an area where the Cu concentrations in-80# (< 177 μm) sized stream sediments highlight the prospective area.

linked to the climatic and physiographic conditions and nature of the commodity being sought. Sizes selected are either coarse (> 250 μm or even > 500 μm) or fine (< 75 μm). In wet climates, fine-grain fraction is preferred due to high chemical weathering in these climates promoting the migration of metals in solution, due to hydromorphic dispersion, which adsorb into the fine-grain clay-sized minerals. In dry semi-arid to arid climates, coarse-sized fractions are preferred (> 250 µm) as mechanical weathering or clastic dispersion is dominant and ore minerals are coarser. Furthermore, dispersion trains are shorter in dry semi-arid climates and longer in wet climates. Often the 125–250 micron range is avoided in dry climatic situations because it mostly reflects as wind-blown quartz.

The use of partial extractions or digest methods are also employed especially in wetter climates to highlight hydromorphically dispersed metals that adsorb onto clay-sized minerals. The partial digests act as chemical sieves where the finer and easily dissolvable mineral fraction is targeted, whereas the sieving of fine-grain sizes (< 75 µm) act as a physical sieve. For gold exploration using drainage geochemistry, fine gold is often preferred due to its uniformity in dispersion and either bulk leach extractable gold (BLEG) where large (> 2 kg) bulk stream samples are reacted with a dilute sodium cyanide solution or fine-sized sediment is sieved (< 75 µm) followed by aqua regia digest and measurement of gold.

4.3.5 Regolith profile geochemistry – soil, laterite, lag, calcrete, saprolite and gossans

In surface environments, the principle of secondary or surficial geochemical dispersion is used to find obscured ore bodies. In many weathered regions, deep and highly weathered profiles have developed to show a vertical zonation. The vertical zonation commences from the slightly weathered rock or saprock to moderately or highly weathered saprolite often but not always capped by a variety of duricrusts overlain by soil. (For an introduction to the terminology of regolith profiles, see Chapter 1 of this volume.) The ore body weathers to provide a secondary geochemical expression (or dispersion halo) in the saprolite and overlying regolith materials including soil. The processes of chemical and physical weathering, erosion and deposition that form regolith and landscapes are the same as those that cause secondary geochemical dispersion. Regolith formation processes cause target and pathfinder elements to disperse during weathering of an ore body thereby providing a wider anomaly size than the ore body itself but at much lower concentrations than the ore body itself. For example, weathering of the ore body will result in a smaller and narrower secondary dispersion in the saprolite but a wider dispersion in the overlying residual laterite, thereby widening the signature from ore (Figure 2.11). There are many studies that have used regolith geochemistry in identifying potential mineralised targets in Australia (Butt et al., 2005).

There are several main mechanisms responsible for dispersing metals in the regolith. The main dispersion mechanism is hydromorphic dispersion, whereby elements in their ionic form disperse due to fluid flow processes of

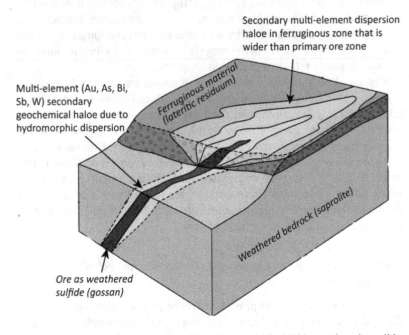

Multi-element (Au, As, Bi, Sb, W) secondary geochemical haloe due to hydromorphic dispersion

Ferruginous material (lateritic residuum)

Secondary multi-element dispersion haloe in ferruginous zone that is wider than primary ore zone

Weathered bedrock (saprolite)

Ore as weathered sulfide (gossan)

Figure 2.11 A generalised secondary dispersion model for highly weathered regolith. *Source:* Modified from R. Smith, CSIRO.

diffusion and advection, both facilitated by downward percolating water and groundwater medium. The chemical processes of dissolution-precipitation, hydrolysis, pH, redox and adsorption all operate to varying degrees within the groundwater and percolating rainwater to mobilise elements. The second mechanism is mechanical dispersion whereby elements are mobilised by physical activity of the regolith material. The regolith material moves or is mechanically transferred vertically and laterally due to bioturbation (bio-mechanical movement due to the activity of animals and roots within the regolith) and near-surface sheet wash and gravity creep processes. The third, but lower intensity dispersion mechanism is via biogenic processes whereby vegetation mobilises metals from the deeper regolith to the surface via their physiological functions. The final mechanism is gaseous dispersion whereby mainly volatile and gaseous elemental forms are transferred to the surface via gaseous diffusion along structural pathways. (This mechanism is restricted to specific elements such as mercury and sulphur that can become volatiles.)

The different regolith sampling mediums are available for geochemical exploration depending on the terrain. This includes stream or drainage samples, soil and surface ferruginous or iron-rich gravels – referred to as lags, ferruginous (or lateritic) duricrust (iron-rich cemented material), calcrete (calcium carbonate cemented soil material) and saprolite (weathered rock). These are all sampled and assayed to provide a clue to the presence of deeper orebody.

Much of the surface geochemical dispersion and sampling is linked to landform evolution of a region because the surface materials are influenced by, or are a direct product of, the manner of landform evolution. Long periods of surface weathering, erosion and deposition of weathered materials result in complex regolith material relationships near the surface, most being linked to the landform in which they occur. Accordingly, conceptual geochemical exploration models have been developed. These models represent links between landforms, underlying regolith materials and surficial dispersion enabling a judicious selection of the appropriate sampling mediums, sampling procedures and sample spacing to be made (Butt et al., 2000; Anand and Butt, 2010). The main landform regimes to distinguish are landforms with residual regolith materials that have evolved from underlying bedrock and whose geochemical signatures provide a robust expression of the deeper rocks (represented in Figure 2.6).

4.3.5.1 SOIL

Soil is probably the most complex medium used in geochemical exploration but is also the most readily available and the most useful as it occurs across a variety of landforms and climatic zones. It generally has distinct horizons with differing properties, each containing a variety of primary and secondary minerals and size fractions. Distinct horizons are not present in all soils and often some are not well developed due to the factors that affect the formation of soils such as nature of the parent material, climatic influences, topography, nature and the organisms living in soils. In exploration geochemistry, the sampling of specific horizons of the soil are preferentially targeted. Several soil classification systems exist and it is important to make a note of the horizon development in soils during a soil survey.

Soil surveys (including duricrust surveys and calcrete surveys) require a sampling pattern to be established covering sampling size and sampling horizon (Figure 2.12). These are best achieved via an orientation survey over known mineralisation in an area. There are a few rules of thumb to bear in mind when designing a soil grid or sampling pattern. When a strike or trend of the potential target is known – for example, from geophysical data or a rock type trend – then the baseline should be parallel to the anticipated strike of the target. Cross lines should be perpendicular to the strike

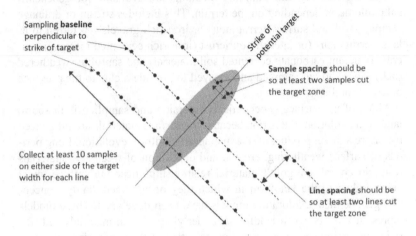

Figure 2.12 Soil sampling parameters for potential target zones. A square grid would be preferred for unknown trend.

of the anomaly or target, and line spacing should be such that at least two cross lines cut the anomaly. Sample spacing should be such that at least two samples are collected from each line crossing the anomaly and at least 10 samples for each line collected from outside the target width. When the trend of the potential target is unknown, then a square grid is preferred at 200 to 400 m spacing.

Sampling soil is done using either a spade or auger. The depth of soil sampled should be dictated by the soil horizon rather than a specific depth so as to maintain a consistent geochemical medium. Soil samples collected are either analysed using a sub-sample of the bulk sample or sieved to select a particular size fraction. Generally the < 2 mm fraction is selected, but in dry, arid and semi-arid environments, a finer (< 177 μm or < 75 μm) or coarser fraction (700–2000 μm) is preferred to avoid the dilution effect of wind-blown sand-sized quartz in the soil (Figure 2.13).

One critical component in deciding to use soil as a geochemical medium is the nature of the parent material, i.e. whether it is residual having evolved from the underlying basement or has been derived from transported materials. In areas of moderate to deeper transported cover (> 10 m), results of soil geochemical surveys need to be carefully interpreted because false positive anomalies are common (Gray et al., 1999) because the upward transport mechanisms of metals are not well understood or clear. On the other hand, selective and/or partial leaches on soil over shallow to moderate transported

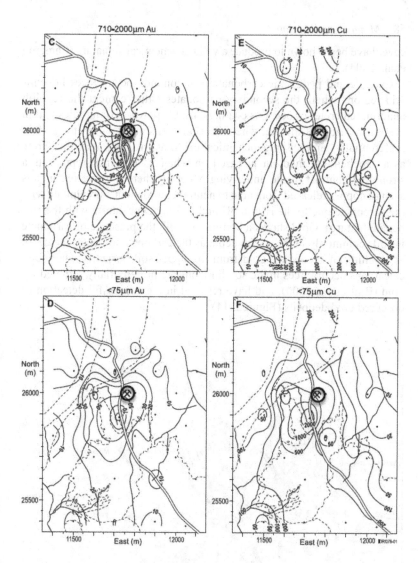

Figure 2.13 Example of soil geochemical survey to detect Fe-oxide Cu-Au deposit Little Eva. Geochemical contours of coarse (710–2000 µm) and fine soil fractions (< 75 µm) show a close relation between the location of the deposit (circle) and the highest values. Although both grain size fractions for both elements highlight the deposit, there are differences in abundances between the grain sizes, for example, anomalous values in finer fractions for Cu are five magnitudes higher than those in coarse fraction. The variation in concentrations between grain sizes highlights the importance of comparing the same sample medium.

Source: After Robertson (2005).

cover have been shown to produce a valid geochemical signature (Cameron et al., 2004).

Calcrete and pedogenic carbonates are common occurrences in semi-arid regions where the secondary carbonates either cement the regolith material (calcrete) or occur as coatings, friable powders or nodules within the soil horizons. Calcrete, where it occurs separately within the soil, is preferentially sampled in geochemical exploration. The value of calcrete as a geochemical medium is based on when its presence represents a strong pH contrast to the underlying material whereby some ore metals such as copper released from acid to neutral environments precipitate with the carbonate due to the rise in pH and thereby concentrate with the carbonate features. Gold shows a correlation with the presence of carbonate in soils commonly in the 1–2 m depth as the carbonate indicates the depth of leaching of secondary gold from the upper soil horizons. Therefore, calcareous occurrences in the soil horizons are preferentially sampled for gold (Butt et al., 2000), and have resulted in the successful detection of obscured gold deposits (Figure 2.14).

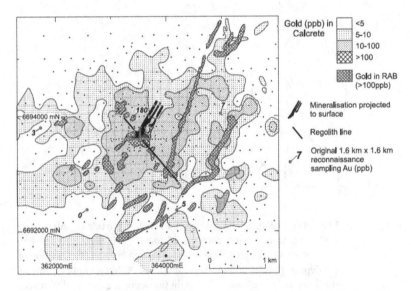

Figure 2.14 Contoured values of gold in soil calcrete on a 100 by 100 m sampling grid over the Challenger Au deposit. Gold anomalous values on a regional scale larger grid (1.6 x 1.6 km) and in Rotary Air Blast (RAB) drilling are also shown.

Source: From Lintern (2005).

4.3.5.2 LAG

Lag is the accumulation of coarse fragments of diverse origins or composition that forms a veneer on the land surface. For exploration purposes, lag generally refers to the high ferruginous material present on the surface and includes pisolites, nodules or fragments of variety of saprolite and mottles. Lag geochemical surveys show broad mineralisation anomalies due to greater mechanical dispersion at the surface (Robertson, 1996), and have been successfully applied in residual cover landforms to detect gold and nickel mineralisation (Figure 2.15).

Lag is sampled by preferentially sweeping the lag material from the surface with a dustpan and brush. Although an un-sieved lag sample collected

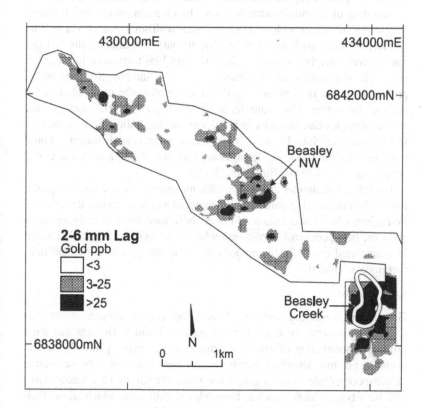

Figure 2.15 Successful geochemical lag sampling example. Contour plot of Au concentrations in lag highlighting the Beasley Creek gold deposit. The main plot is of regional lag survey and the inset from a detailed survey of the area marked Beasley NW.

Source: From Robertson and Carver (2005).

is satisfactory, coarse (10–50 mm) and fine lag (0.5–10 mm) fractions are more effective (Robertson, 1996). Magnetic lag comprises maghemite, and can be preferentially sampled using a hand magnet, although studies have shown it not to provide a significant advantage over whole lag samples.

4.3.5.3 LATERITIC RESIDUUM

Lateritic residuum is the upper part of the iron-rich zone of a highly weathered profile (see Chapter 1 for more) and consists of iron-rich unconsolidated gravels and iron-rich duricrusts that have been cemented by mainly fine iron oxide and hydroxide minerals. Lateritic geochemical surveys have shown to be successful in defining large kilometer-wide anomalies from base metal, gold and pegmatite deposits (Smith and Pedrix, 1983; Anand and Butt, 2010).

Sampling of lateritic residuum is done by chip sampling when outcropping at the surface or with an auger for areas with soil cover or via drilling for deeper zones such as when they occur buried. Sampling the ferruginous gravel materials should be done without bias to type of fabric, such as nodule or pisoliths, and a minimum of a 1 kg sample should be taken, preferably over a 10 m radius. The depth of sampling can be dictated by the scale of the survey. For example, in reconnaissance lateritic surveys, the surface samples can be taken at a 1–3 km spacing collecting near surface ferruginous materials (Figure 2.16), whereas prospect scale surveys should preferentially sample lower ferruginous horizons of the duricrust at closer sample spacing (400 m) (Butt et al., 2000).

The critical parameter in using lateritic materials in geochemical exploration is their underlying parent source material and to an extent the climate. Ferruginous loose and indurated materials formed from underlying rocks show the best geochemical response whereas ferruginous materials formed in transported cover (sediments) do not. The interpretation of landform, therefore, is crucial.

4.3.5.4 GOSSANS

Gossans are weathered varieties of rocks that contain substantial amounts of sulphide minerals such as matrix or massive sulfides. Gossans that are a product of weathering of iron-rich sulfides such as pyrite generally contain hematite, goethite, jarosite, whereas gossans that are a product of weathering of iron-poor sulfides such as galena and sphalerite tend to be siliceous. Gossans have typical fabrics such as boxwork and colloform, which allows their discrimination from other ferruginous regolith. Multi-element geochemistry of gossans is required as both target and pathfinder elements can be variably depleted in gossans due to intense leaching. The multi-element suites for various mineralisation types were tabulated by Taylor and Thornber (1992) and the use of multivariate statistical analysis techniques such as stepwise discriminant

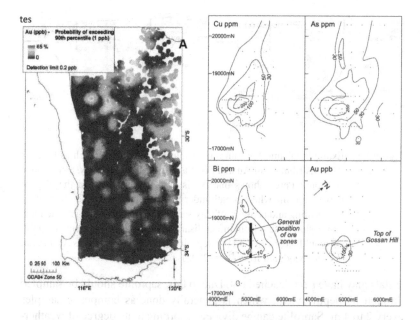

Figure 2.16 Application of laterite geochemistry in deeply weathered terrains. A. Kriged values of Au in laterite samples collected at 3 km sample spacings across the Yilgarn Craton showing higher values approximate in locations with Au mineralisation (from Cornelius et al., 2008). B. Multi-element (Cu, As, Bi, Au) concentration contour values in laterite at Gossan Hill, WA, showing variable dispersion of different elements from mineralised gossans (Smith and Pedrix, 1983). Note that different elements show variable spatial dispersion as noted from their values with Au being restricted to ore zone and Cu and As showing the widest dispersion.

analysis are best suited to gossan geochemistry. The identification of gossans near or at the surface has led to the detection of several base metal deposits such as Broken Hill and Mt. Isa. Although gossans are less likely to be detected close to surface nowadays in exploration surveys, gossans may still be recognised in moderate to deeper transported cover via drilling.

4.3.5.5 SAPROLITE

Saprolite, or moderately to highly weathered rock, is a common geochemical sampling medium being sampled in shallow drilling programs, as it is likely to reflect the chemistry of the bedrock it is derived from. In saprolite, however, depending on conditions of weathering, some ore and pathfinder

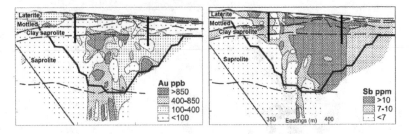

Figure 2.17 Example of saprolite sampling using shallow drilling such as RAB or AC. Distribution of Au and Sb in a section across the Mystery deposit at Mount Percy shows that Au is depleted (leached) in the clay or upper saprolite whereas pathfinder Sb is not depleted. Therefore, shallow drilling and analysis of clay saprolite may not detect geochemical expression arising from mineralisation whereas a multi-element analysis would have detected the deposit via pathfinders such as As and Sb. *Source*: Modified from Butt (2005).

metals may have been leached out. Due to this, saprolite should be sampled through its depth extent, which is generally done as composite samples every 2 to 4 m. Saprolite can be divided according to its degree of weathering or oxidation into upper or lower saprolite or fine or coarse saprolite, and these zones can also be specifically sampled. An end of hole (EOH) sample is taken from the least weathered rock or saprock.

Interpreting saprolite geochemistry needs to consider its weathering degree, the spatial extent of geochemical halo or dispersion and leaching of specific elements. The spatial extent of secondary dispersion halo is likely to be narrow in saprolite and saprock (Figure 2.10) so there needs to be careful consideration of sample spacing in order to detect a dispersion halo. The weathering degree affects the leaching of elements with soluble elements often being completely leached. Depending on geochemical conditions, some ore metals may be also leached from specific parts of the saprolite such as gold in saline condition oxidizing conditions (Figure 2.16). Saprolite sampling and analysis via surface outcrops and shallow drilling should, therefore, consider the intensity of weathering and the element suite required to highlight any leaching of specific ore metals, and multi-element analysis should be conducted rather than single element.

4.3.5.6 TRANSPORTED COVER

Transported cover refers to the exotic Earth material unconformably overlying the fresh or weathered mineralised basement, and is therefore post-mineralisation. It is mostly sedimentary in origin, dominated by unconsolidated to cemented surficial sediments deposited over basement rocks due to aeolian, fluvial, glacial, gravity and lacustrine processes. However, the term *transported cover* increasingly is being also used to refer to lithified basin sedimentary rocks (see Figure 2.3). Transported cover is also referred to as simply 'cover' or 'transported overburden.'

Across much of the Southern Hemisphere continents, shallow transported cover is a combination of fluvial, aeolian and colluvial sediments whereas across the Northern Hemisphere it is predominantly glacial in origin. The primary method to explore in these transported covered regions is geophysics, although geochemical methods are also used.

The application of selective or partial leaches on soils developed over transported cover sediments is commonly used (see Figure 2.9 for a list of selective leaches). The rationale of using selective leaches rather than whole soil or a particular particle size of soil is that selective leach targets specific loosely bound ions within the soil. The premise is that the loosely bound component represents recently transported labile or 'mobile' ions that have migrated upwards through transported cover. The selective leach is able to highlight this labile or exogenic component from the background endogenic component of an element that is strongly 'bound' to soil matrix because it was present as part of the cover sediments. Although several case studies demonstrating the effectiveness of selective leaches in highlighting buried mineralisation have been noted (Cameron et al., 2004), the poor understanding of the mechanisms capable to disperse metals upwards through cover has been a limiting factor in the effective interpretation of the selective leach survey data and therefore the application of selective leaches. Nevertheless, selective leaches are resorted to across many transported covered terrains.

There are other methods besides selective leaches that are used. Across the Northern Hemisphere continents which were covered by Quaternary ice sheets, glacial sediments commonly overlie basement mineralised rocks. As the glacier advances and scours the underlying rocks, it carries rock materials including mineralised grains. When glaciers melt, the scoured rock debris is deposited as thin to thick sediments. The common glacial sediment is till, which is a combination of poorly sorted clay, silt sand and gravel. The debris deposited by a glacier forms a flame to ribbon shaped three-dimensional lens of sediment starting from the source and this debris lens is referred to as dispersion train. If the source is mineralised, then the dispersion train represents a much larger dispersion plume than the source. The concentration of an element or indicator mineral when plotted against distance from the source shows an exponential decline down ice direction within the till, with the highest concentration near the source (DiLabio and Coker, 1987; McClenaghnan and Paulen, 2017), and the decay is similar to that observed for drainage sediments. The geometry of the dispersion train is governed by the ice flow directions, erodibility of the source rock, topography of the area and post-depositional processes (DiLabio and Coker, 1987). The concept of dispersion train and its variations according to the number of ice flow phases is shown in Figure 2.18. By systematically sampling the till and analysing its geochemistry and indicator minerals, the concentration of the tills constituents can be used to first identify the dispersion train and then to trace back 'up-ice' to the source. The entire process of using till geochemistry and lithology to trace the source is referred to as drift prospecting.

Till is generally collected on a grid pattern. The depth of till sampling depends on the nature of the stratigraphy of the till; single thin units (< 2 m) are easy to sample via a near-surface sample, whereas multiple sheets of till require sampling all the sheets and therefore a drilling method, as dispersion trains may occur in one or more layers of till. Surface till samples in simple stratigraphy are taken using a shovel or mechanical excavator whereas deeper samples from multiple till layers are sampled using augers or rotary drills (preferably rotasonic drills that limit loss of fines). The till-sized fraction commonly analysed is the clay + silt fraction (< 0.063) because many of the indicator minerals are within this size range and the clay minerals found in clay-sized fraction adsorb and incorporate elements released due to weathering of the till and mineralised minerals. The sand-sized fraction is avoided or excluded as it hosts abundant quartz and feldspar grains that act as a dilutant, similar to that in arid and semi-arid regions due to wind.

Besides applying selective leaches to soils developed within transported cover materials, drilling is the method of choice to target sub-surface units especially in regions where stratigraphy of the sediments is unknown or varies considerably over short distances. An increasingly used sample media identified and sampled through drilling is referred to as interface – a sample taken across an unconformity (Robertson, 2001). Interface sampling was initially proposed as sampling across residual basement (saprolite, saprock, mottled saprolite) but due to multiple unconformities being present in deeper cover sediments, it is also applied across one to two unconformities e.g. basement and Permian sediments, Permian sediments and Cenozoic sediments. The premise of selectively sampling interfaces is that they represent older land surfaces and therefore wider geochemical dispersion is likely to have been occurred and retained across these boundaries due to a combination of hydromorphic and colluvial or mechanical dispersion. Interface sampling involves taking a sample above the unconformity and then immediately below it. In rotary drilling methods that do not extract a core, it is often difficult to identify and pick unconformities accurately, and so sampling interfaces are often subjective and not accurate. The common method used is to sample every 2 or 4 m composites of regolith and rock cuttings downhole and assay them.

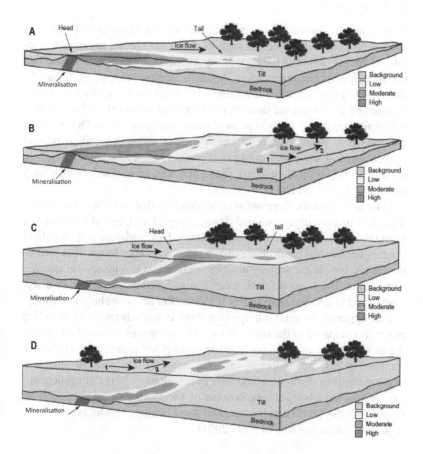

Figure 2.18 Conceptual models of dispersion trains along ice flow directions. A.
A simple case where the close to surface mineralisation as source is
dispersed via single ice flow event leaving a contrasting geochemical
or mineral signature from tail to head of the train. C. A case where there
is a thick till cover and the dispersion train commences from the buried
source but is close to the surface at the tail i.e. down ice. B and D are
modified cases of A and C respectively whereby the original train is
diluted and offset by second ice flow events.

Source: From McClenaghnan and Paulen (2017).

4.3.6 Vegetation

Vegetation can also be used in mineral exploration. Plants absorb elements (nutrients) from soil and deeper regolith and some species accumulate specific elements (indicator plants). Many elements are selectively absorbed (calcium, potassium, magnesium, sodium, sulphur, copper, iron, molybdenum, selenium and zinc) from the sub-surface by the root network and transported above ground through the roots and stems driven by capillary action and transpiration (as in plant physiological processes). The absorbed elements may reflect deeper or below-ground anomalies and so present biogeochemical anomalies in plants (Brooks et al., 1995; Dunn, 2007). In addition to the essential elements, plants may also take up toxic metals such as Ag, As, Au, Cr, Ni and U.

In sampling plants, there are several variables that can affect the results. The plant species, organ sampled (leaves, twigs, bark), age of the plant and even the season in which the plant is sampled are variables that can affect the biogeochemical result. Ease of sampling is another factor. It is recommended to sample the actively growing ends of leaves with several samples being taken from around each tree. The sampled leaves are then dried, ground and milled and analysed by ICP-MS after acid dissolution or ashing.

It is important to establish whether there is any advantage to sampling plants as compared to the adjacent soil. The advantage of sampling soils is the small number of variables (such as horizon and grain size), whereas in sampling plants, the number of variables increases. The main advantage of using plants is in shallow transported cover regions (< 30 m). Biogeochemical surveys have had success in areas of transported cover, especially in Canada (Dunn, 2007) and have shown some positive results even in drier regions of Australia (Anand et al., 2007).

4.3.7 Groundwater

Groundwater occurs in unconfined (mostly in regolith) and confined aquifers (in deeper cover regions). Groundwater geochemical surveys work on the principle of detecting dissolved metals and isotopes that are dispersed in the groundwater due to the weathering of deposits or from secondary dispersion haloes surrounding deposits in the regolith. By sampling the groundwater and analysing for target and pathfinder elements, the concentrations of selected elements may narrow the area of interest. From all of the geochemical methods that are available, groundwater surveys theoretically have the most potential to detect mineralisation concealed under moderate to deep cover.

Groundwater is generally sampled from exploration drillholes, water bores, windmills and wells (Gray et al., 2011). Ideally, the drillholes and wells need to be cased and screened. However, for exploration, such equipped bores are rare and most groundwater surveys are done with open drillholes and wells. Water is acquired by bailing from a bore or drillhole or from a windmill. The sample then needs to be immediately filtered through a 0.45 μm filter and stored in clean HDPE plastic bottles in a cool place. The pH, temperature, conductivity (EC) and oxidation/reduction potential (ORP) are measured immediately on exposure to air. Electrodes used to measure the water parameters need to be calibrated regularly. Gold concentrations in groundwater are extremely low (in the ng/L range) and, therefore, gold needs to be concentrated before analytical measurement. This is achieved by placing activated carbon sachets in 1 L of extracted groundwater as the carbon adsorbs and concentrates the gold from the water. Subsequently, the gold concentration in the carbon is measured. Unlike solid samples, the water samples need to be analysed for anions using ion chromatography (IC) and cations using ICP-OES or ICP-MS. Isotopes are increasingly being measured in groundwaters and the main methods to measure radiogenic isotopes (strontium and lead) is thermal ionisation mass spectrometry (TIMS), but increasingly multi-collector ICP-MS instruments are also used. Stable isotopes in groundwater are measured via the Isotope Ratio Mass Spectroscopy (IRMS).

Groundwater concentrations should be first corrected for charge-balance errors and then plotted to highlight any anomalies, such as lead Pb or zinc Zn for VMS deposits (Leybourne and Cameron, 2010). Further calculations of saturation indices, element ratios and indices have shown promise in distinguishing bedrocks and mineralised rocks from barren ones (Giblin, 2003; Gray et al., 2016). For example, multi-element indices using pH, Eh, Fe, Mn Mo, Li, Ni and Co demonstrated the value of hydrogeochemistry to separate zones of Ni-sulfide mineralisation from surrounding regional unmineralised ultramafic rocks (Gray et al., 2016). Isotopes in groundwater have also shown promise with (S, O, Pb and Sr) isotope variations being interpreted to indicate contribution from oxidising base metal mineralisation (de Caritat et al., 2005).

However, several chemical and physical parameters – such as pH (acidic vs neutral to alkaline), redox states (reducing vs oxidising) and conductivity (fresh vs saline) of the groundwater – influence the interpretation of hydrogeochemical data. Acidic, compared to neutral to alkaline groundwaters, affect the mobility of metals, including precious, base metals and pathfinders (cations and oxyanions). For example, As and Se provided a widespread anomaly around the concealed Spence deposit in Chile as these metals formed oxyanions and, therefore, travelled farther down-gradient in neutral to slightly alkaline groundwater (Leybourne and Cameron, 2010). Another example is where saline, oxidising groundwaters contain more Au in solution due to complexing with chloride (Gray, 2001). Furthermore, mixing of groundwaters either from different aquifers or from surface waters can dilute anomalies arising from deeper mineralisation. Isotopic analysis of the groundwaters can be used to estimate the mixing of groundwaters from different sources. Interpretation of hydrogeochemical survey data needs to consider groundwater mixing in addition to pH, salinity and redox.

4.3.8 Lithogeochemistry

The use of geochemical data from the primary dispersion environment whereby enrichment or depletion zones of mineralised or hydrothermally altered rocks are detected is known as lithogeochemistry (Govett, 1983). As most of the exploration is currently conducted under some variation of cover, much of the lithogeochemistry data is derived from drilling. Lithogeochemistry applied to mineral exploration is mostly used in three ways:

- To identify barren versus fertile rocks and also rock types that have been altered (mainly igneous) using immobile trace elements such Zr, Nb, Ti and Sc;
- To identify hydrothermal alteration using variations in major elements (Na, K, Al, Ca, Mg) that reflect mineralogical changes in altered rocks (Figure 2.19), which aids in narrowing the ore zone;
- To establish pathfinder element patterns to provide a direction to the ore. An example of using major and pathfinder element geochemistry and patterns to pinpoint the location of alteration types and ore is shown in Figure 2.16, where different elements are enriched in different type and intensity altered zones of a hydrothermally altered ore system.

Identifying rock types that have been hydrothermally altered or weathered (saprolite and saprock in drillholes), is best done using a combination of scatter plots of immobile elements (i.e. elements that are known not to be lost or gained during hydrothermal and chemical weathering processes). Such elements are zircon, titanium, niobium and scandium. For example, plots of Zr vs Ti have been used to identify volcanic rocks from each other in saprolite samples (Hallberg, 1984) including Ti vs Nb, and scatter plots of Sc vs Cr for ultamafics. V vs Ti for mafic rocks, and a combination of plots of the distinguishing elements will allow separation of altered rocks.

Lithogeochemistry is increasingly being used to identify and classify hydrothermal alteration. To apply lithogeochemistry to alteration requires an understanding of the different hydrothermal alteration types, which in turn are a function of the diagnostic minerals or mineral assemblages and chemical changes (Pirajno, 2010). The alteration intensity is subjective and commonly described as weak, moderate, strong or intense depending on the preservation of the original rock and minerals, which is a function of the chemical changes in hydrothermal solutions with temperature. The chemical changes effected to the original rock by way of changing K, Na, Ca, Fe (enrichment or depletion) are measured and potentially quantified by the lithogeochemistry.

Identifying hydrothermally altered zones in drill cuttings (reverse circulation drillholes) and diamond core is used to interpret the direction to ore that is referred to as alteration vector, especially in deeper transported cover areas (Figure 2.19). The type and intensity of hydrothermal alteration can be recognised using molar ratios of major elements. The two common molar ratios recommended are Pearce Element Ratios (PER), which use an immobile element such as Al or Ti, and General Element Ratios (GER), which do not use a conserved element as a denominator (Whitbread, 2002). The scatter plots are so designed to place minerals that are found in the background or unaltered rocks farther away from the minerals that represent hydrothermal alteration. For example, a common GER plot would have K/Al against Na/Al (Figure 2.19). In such a plot using K/Al and Na/Al molar ratios, the unaltered rocks should have a composition that plots closer to the center, whereas sericite (muscovite) rich sample will plot closer to the muscovite zone (0,1/3), while potassic altered samples will plot closer to the biotite apex. Transferring the down-hole samples that display the types of alterations from the GER plots in 3D software allows visualisation of the scale and trends represented in the alteration systems and thereby enables one to identify the direction of the main ore.

Another method to illustrate complex geochemical variation due to hydrothermal alteration is to combine the key elements associated with alteration in a geochemical alteration index such as Na, K, Mg, Fe and Ca. The elements are generally ratioed to maximise the effects of alteration. For example, enriched elements may occur in the divisor with the depleted elements. A commonly used AI is the Ishikawa Alteration Index after its originator which helps quantify the intensity of sericite and chlorite alteration in volcanic rocks, and the chlorite-carbonate-pyrite index (CCPI), that enables to measure the intensity of formation of Mg-Fe chlorite after albite and K-feldspar in volcanic rocks as well as the carbonate alteration and pyrite enrichment (Large et al., 2001). The two indexes are:

$$AI = \frac{100(K_2O + MgO)}{(K_2O + MgO + Na_2O + CaO)}$$

$$CCPI = \frac{100(MgO + FeO)}{(MgO + Na_2O + FeO + K_2O)}$$

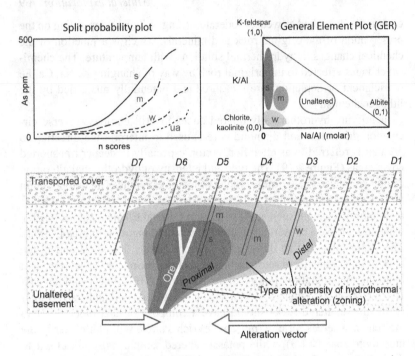

Figure 2.19 The concept of using lithogeochemistry under cover as a vector to mineralisation. Lower cross-section illustrates the style and intensity of hydrothermal zoning around an ore deposit with strong alteration (s) closer to ore (proximal) and weaker alteration (w) commonly represented by chlorite (propylitic) further from ore (distal). Analysis of down-hole data as either end of hole or meter composites, when visualised on a general element ratio plots (GER) (top right) can identify the intensity or class of alteration of each sample (e.g. strong, moderate, weak or un-altered). In the GER example shown, plot of K/Al vs Na/Al molar ratios depict the end members represented by molar ratio of minerals within a hydrothermal alteration system. The approximate location of molar ratios of unaltered rocks is shown which when altered will shift rock composition to different mineral end points. Furthermore, target or pathfinder element probability plots of the data can reveal subtle variations in concentration depending on distance from ore. The GER and pathfinder concentration assessment can assist in identifying the alteration vector towards mineralisation. For example, in the lower cross-section, drillholes D1 and D2 do not detect any alteration, whereas D3 and D7 detect weak alteration, D4 detects moderate alteration and D5 and D6 detect strong and moderate alteration.

5 Drilling in exploration

Let us view a scenario where either a combination of geological, geochemical and geophysical methods or a single discipline has been used to define a target. For example, geological mapping has identified an alteration zone near surface. In erosional terrains, follow-up stream sediment geochemistry may have been used to pinpoint targets. In residual regolith terranes, magnetics may have been first used to narrow an area over which a soil geochemical anomaly was defined. Possibly, in regions of deeper transported cover, regional interpretation of electromagnetic data may have defined a deeper conductive target. Irrespective of the method used to define the target (i.e. geochemical and/or geophysical), the only way to test the target is to drill it and subsequently use the drill samples to identify geology and also to provide geological and geochemical indicators of mineralisation. An exploration geologist needs to use the type of drilling suitable to the stage of exploration to test the target.

There are different rotary drilling methods available to test targets. The drilling methods – Rotary Air Blast (RAB), Air-core (AC), Reverse Circulation (RC), Diamond Drilling (DD) and Tube Coil (TC) – all have their pros and cons, and are used based on the stage of the exploration program and aim of the project. Although drilling methods do not form part of the 'science' of exploration, they are critical to the quality of the data produced and the reporting of that data via the JORC or other acceptable codes (e.g. The Samrec Code, N143–101). Importantly, they are the main tool available to convert mineralised targets into inferred resources and subsequently further into indicated resources and reserves.

In erosional terrains and residual terrains, where the regolith is thin and weathered rock close to surface, digging pits and trenches (referred to as costeans) – either by hand or by using excavators – can provide valuable information on the lithology, structure and assays, and compliment RAB or AC drilling (Marjoribanks, 2010).

5.1 Rotary Air Blast (RAB)

In the operation of RAB drilling, high-pressure air is blown down inside the hollow of a drill pipe through a rotating tungsten bit attached to the end of the drill pipe. The rotating bit grinds through the regolith and pressurised air released from the bit at the cutting end lifts the cuttings from the bit face along hole walls to the surface (Figure 2.20). This method is only good to drill through 'softer material' (i.e. soft regolith), and cannot penetrate fresh rock. The regolith cuttings lifted to the surface pass through a cyclone where the air is separated from the samples. This method is the cheapest and fastest form

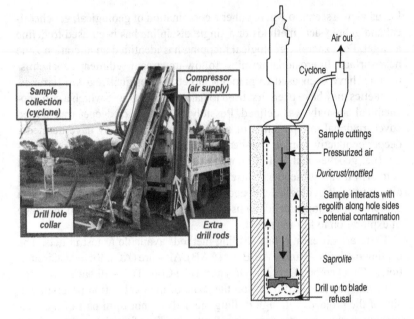

Figure 2.20 RAB drill rig showing the main parts of the rig (Image: Arinooka Drilling). A schematic view of the operation of RAB drilling (right).

of exploratory drilling and provides a large sample volume. The disadvantages are that the samples move along hole walls and therefore interact with the regolith along the wall, resulting in potential geochemical contamination. Therefore, if a gold-bearing vein is encountered at 10 m depth, all samples from below that depth will interact with the vein and register small gold values in the assay of the sample even though these samples contain no gold. Furthermore, sample depth determined by this drilling method is not accurate and as the method can only penetrate 'soft' material, finer cuttings can be lost in cavities present along the drill-hole wall. It is, however, an ideal method for rapid 'screening' of an exploration target area or prospect. The method is used only in the initial exploration stage and is not suitable for defining the orebody or resource evaluation. The method's operational principles are the same for blast hole drilling conducted during open pit mining operations.

5.2 Air Core (AC)

In Air Core (AC) drilling, the drill pipe setup is different to RAB drilling. The drill rods are hollow and have an inner tube inside the rod. Unlike RAB, the drill cuttings are lifted through the inner tube by compressed air. The cuttings, therefore, do not interact with hole walls and maintain geochemical integrity. The drill bit is hardened steel or tungsten and has three blades that assist in cutting the rock. The drill bit is capable of drilling a small piece of core at the bottom of the hole. The advantage of AC over RAB drilling is that sample integrity is maintained and pieces of core can be obtained. Its limitation is that it, too, cannot penetrate harder or fresh rock.

5.3 Reverse Circulation (RC)

In Reverse Circulation (RC) drilling, dual barrel/tubes are used. These consist of outer and inner tubes, similar to the AC drilling setup. Compressed air or water is driven down the outer or annular space to the tungsten drill bit (round buttons) that cuts through fresh hard rock. A percussion hammer is commonly used to penetrate harder formations. The drill cuttings are returned to the surface through inner rods for collection through a cyclone that uses centrifugal force to separate the solids from the airflow (Figure 2.21). This drilling method requires high air pressures and large rigs. The advantages of RC drilling include good depth penetration (~500 m deep) with the capability to drill through regolith and fresh rock. Because rock chips return to surface through the inner pipe, there is minimal contamination and good sample integrity. The disadvantage of RC drilling, compared to RAB/AC drilling, is the higher cost. Compared to diamond drilling, RC drilling only provides rock cuttings and so no continuous rock information and structural data can be determined. The RC drilling method is the most common drilling method used from initial to advanced stages of exploration.

The geological data acquired from AC or RC cuttings is used to identify the rock type and, to an extent, mineralogy and alteration. Assays may be carried out on 1 m, 2m or 4 m composites. The samples are logged by geologists and a small wet sieved sample can be stored in chip trays.

5.4 Diamond Drilling (DD)

Diamond drilling consists of a series of rods with a diamond-impregnated bit that produces a solid core of rock. The drill bit is lubricated with water or muds (special drilling fluids). The core barrel is subsequently winched to the surface at the end of each run.

Most diamond holes (and RC holes) are designed as inclined holes to provide maximum information from steeply dipping stratigraphy. The dip and azimuth are recorded for each hole. Down-hole surveys are carried out to measure dip and azimuth at specific depths using a specialised down-hole survey camera. Wedge (or daughter holes) can be established in diamond

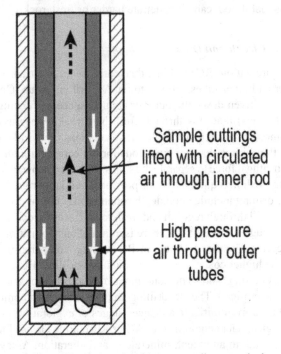

Sample returns through inner rods; contamination free

Sample cuttings lifted with circulated air through inner rod

High pressure air through outer tubes

Figure 2.21 A Reverse Circulation rig with a cone splitter attached to the cyclone. A schematic view of the operation of RC rig.

holes to help define specific ore extensions of the known mineralisation without the expense of drilling new holes from the surface (Figure 2.22). Diamond holes can be drilled using different core diameters, such as BQ (27 mm diameter), NQ (47.6 mm), HQ (63.5 mm) and PQ (84 mm). Diamond drilling can also be conducted underground from different levels to explore for the extensions of existing ore bodies or to define ore dimensions clearly.

One of the main uses of diamond core as compared to RC/AC is to study geological structure (Marjoribanks, 2010). To correctly interpret common structures in core, the cylinder of the core needs to be correctly oriented in addition to the azimuth and dip being carefully measured. The orientation

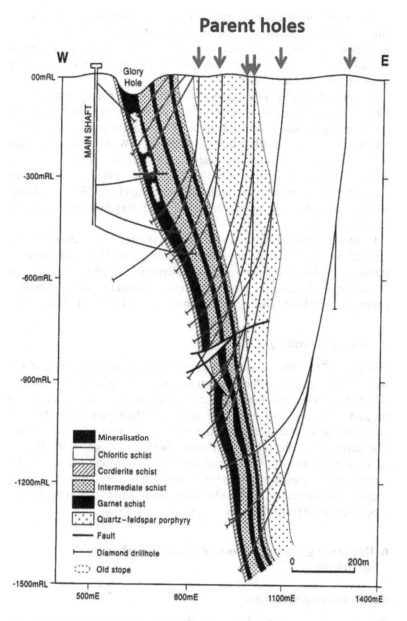

Figure 2.22 An example of deliberate deviation of diamond holes from a single parent hole. The controlled deflection of deep diamond holes allows definition of multiple intersections of mineralisation from six parent holes.

Source: Big Bell Gold Mine, WA, after Handley and Cary (1990).

of the face of the core is measured using several methods such as the older down-hole spear, mechanical systems such as Ballmark[R] and Ezy-Mark[TM], and newer electronics ones such as Reflex[TM]. Irrespective of the method used to orient the core, it is imperative to orient the core correctly and mark the orientation line accurately.

Diamond drillholes are laid out in core trays and photographed before further processing, such as cutting for assays. Newer techniques allow the entire core to be assayed (Minalyzer XRF), and reflectance spectra in the visible near infrared (VNIR), shortwave infrared (SWIR) and thermal infrared can be recorded using instruments such as HyLogger (Schodlok et al., 2016) and CoreScan (www.corescan.com.au/). Reflectance spectra facilitate near continuous (1mm spatial resolution) mapping of specific minerals and mineral assemblages, especially those indicative of hydrothermal alteration (Cudahy, 2016).

Diamond drilling is the most effective but also the most expensive drilling method. Due to its high cost and slow drilling rate, it is used in advanced stages of exploration and resource development. Its effectiveness is the advantage of providing the precise and best geological information to analyse an ore body for detailed geological, metallurgical and geotechnical tests.

5.5   Coil Tube drilling (CT)

Coil Tube (CT) drilling uses a continuous, flexible steel coil instead of drill rods or tubes, thereby reducing the time and cost of adding and retrieving drill rods used in other methods. The CT drilling method uses a motor at the end of the drill coil string to penetrate the formation, rather than the conventional method of rotating the entire drill string from the surface. The advantage of the CT method over other conventional methods (AC/RC) is the elimination of individual drill rods. CT drilling technology currently being developed may incorporate top-of-hole sensing that will enable accurate measurements of geochemistry, mineralogy and petrophysical properties of the rocks (Hillis et al., 2014).

6   Resourcing the ore deposit, resource database and modelling

6.1   Resource development

The aim of resource development is to demarcate the spatial limits (i.e. geometry) and grades of mineralisation to allow definition of Indicated Resources according to the JORC Code (Chapter 3). The orientation of and continuity of the depth of mineralisation also needs to be determined.

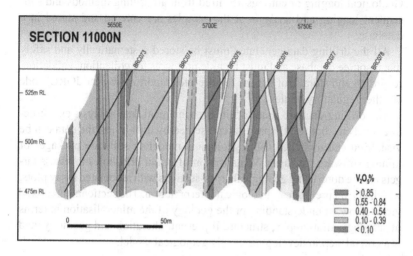

Figure 2.23 Interpreted mineralised cross-section constructed from angled RC drilling and assays. It is necessary to at least intersect mineralised bands so as to define their orientation and continuity.

Source: Image: Reed Resources.

Diamond core and reverse circulation drilling methods are used at the resource development stage where drilling is conducted based on a regular pattern and hole spacing that is appropriate for the type of mineralisation being defined. For example, for relatively homogeneous ore types such as bedded iron deposits and bauxite deposits, wider hole spacings are justifiable, whereas for vein gold systems, where the thickness of veins and widely fluctuating grade values are the norm, much closer drill spacing will be necessary. Depending on the deposit type and orientation of mineralisation, at least two intercepts for most mineralised beds or bands should be made, as it enables the interpretation of the continuity and geometry of the mineralisation (Figure 2.23).

6.2 Resource database

Exploratory and resource development drilling generates a large amount of information. Drilling will result in the acquisition of rock cuttings or core that need to be sampled systematically for a number of purposes for the resource project. Many drill rigs also collect down-hole data using televiewers and down-hole 'logging' tools, such as gamma ray instruments.

Geological logging of cuttings acquired from all drilling methods and subsequent logging and analysis of the samples are entered and stored in a database.

All the drilling data generated must be stored systematically and safely in a resource database and must include protocols for collection, validation, recording and storage of data as required for reporting (by the JORC Code or other regulatory bodies).

The drilling data are used for lithogeochemical data analysis and structural analysis, but as 3D space is represented by drill-holes, the data can be visualised and modelled using geological modeling software packages or mining software packages. Visualisation of initial drill-holes to assess targets can be done using geological cross-sections with associated assay plots, which allows preliminary geological interpretation. The sectional integrated views allow an understanding of the geology of the mineralisation in terms of lithology, stratigraphy, structure if present, alteration and geometry, and is a crucial step in developing the 3D geological model.

6.3 Geological modelling

Extensive drilling of mineralised intersections is used to produce a 3D data set of down-hole geological information. This data can be visualised and modelled either using wireframing techniques or other explicit modelling techniques where logged geology (and other variables such as assay) are subsequently digitised, triangulated and validated to produce a 3D model. In contrast, 3D models of mineralisation and geology are created by the implicit modelling technique where the radial basis function (RBF) interpolation is applied to carefully logged and labelled down-hole data to rapidly create a model, that needs to be validated, but the model honours the drill-hole data (Cowan et al., 2002). The implicit modelling technique is gradually becoming the method of choice given its time efficiencies and ability to cater for faster updates on new drilling data. The advantage of constructing geological models in 3D space is that it allows the geologist to determine the orientation and geometry of the mineralisation, which then allows planning of further drilling for resource development.

References

Anand, R.R. and Butt, C.R.M. (2010). A guide for mineral exploration through the regolith in the Yilgarn Craton, Western Australia. *Australian Journal of Earth Sciences*, 57(8), 1014–1114. DOI:10.1080/08120099.2010.522823

Anand, R. R., Cornelius, M., and Phang, C. (2007). Use of vegetation and soil in mineral exploration in areas of transported overburden, Yilgarn Craton, WA: a contribution towards understanding metal transportation processes. *Geochemistry: Exploration, Environment, Analysis*, 7, 267–288.

Arndt, N.T., Fontbote, L., Hedenquist, J.W., Kesler, S.E., Thompson, J.F.H., and Wood, D.G. (2017). Future global mineral resources. *Geochemical Perspectives*, 6, 1–166.

Brooks, R.R., Dunn, C.E., and Hall, G.E.M. (1995). *Biological Systems in Mineral Exploration and Processing*. Ellis Harwood, New York.

Butt, C.R.M. (2005). Mystery zone gold deposit. In *Regolith Expressions of Ore Systems* (Compliers:C.R.M. Butt, M. Cornelius, K.M. Scott, and I. Robertson). http://crcleme.org.au/Pubs/Monographs/RegExpOre.html.

Butt, C.R.M., Anand, R., and Lintern, M.J. (2000). Evolution of regoliths and landscapes in deeply weathered terrain: Implications for geochemical exploration. *Ore Geology Reviews*, 167–183.

Butt, C.R.M., Cornelius, M., Robertson, I.D.M., and Scott, K. (Eds.) (2005). Regolith expression of ore systems: A compilation of geochemical case histories and conceptual models. *CRC LEME*. http://crcleme.org.au/Pubs/Monographs/RegExpOre.html.

Cameron, E.M., Hamilton, S.M., Leybourne, M.I., Hall, G.E.M., and Mcclenaghan, M.B. (2004). Finding deeply buried deposits using geochemistry. *Geochemistry: Exploration, Environment, Analysis*, 4, 7–32.

Cornelius, M., Robertson, I.D.M., Cornelius, A.J., and Morris, P.A. (2008). Geochemical mapping of the deeply weathered western Yilgarn Craton of WA, using laterite geochemistry. *Geochemistry: Exploration, Environment, Analysis*, 8, 241–254.

Cowan, E.J., et al. (2002). Rapid geological modelling. In *Applied Structural Geology for Mineral Exploration and Mining, International Symposium Abstract Vol* (Ed. S. Vernecombe), AIG Bulletin No. 36, 3639–3641.

Cox, D.P., and Singer, D.A. (1986). Mineral deposit models. *U.S. Geological Survey Bulletin*, 1693, 379.

Cudahy, T. (2016). Mineral mapping for exploration: An Australian journey of evolving spectral sensing technologies and industry collaboration. *Geosciences*, 6, 52.

De Caritat, P., Kirste, D., Carr, G., and McCulloch, M. (2005). Groundwater in the Broken Hill region, Australia: Recognising interaction with bedrock and mineralisation using S, Sr and Pb isotopes. *Applied Geochemistry*, 20, 767–787.

Dentith, M. and Mudge, S.T. (2014). *Geophysics for the Mineral Exploration Geoscientist*. Cambridge University Press, Cambridge, 454.

DiLabio, R.N.W. and Coker, W.B. (1987). Mineral exploration in glaciated terrain using till geochemistry. *Episodes*, 10, 32–34.

Dunn, C. (2007). Biogeochemistry in mineral exploration. In *Handbook of Exploration and Environmental Geochemistry*, Elsevier, 480.

Giblin, A.M. (2003). Groundwater geochemistry from some major gold deposits and exploration targets in northern Victoria and southern NSW. In *Victoria Undercover, Benalla 2002* (Eds. G.N. Phillips and K.S. Ely), CSIRO Publishing, 101–105.

Govett, G.J.S (1983). Rock Geochemistry in Mineral Exploration. *Handbook of Exploration Geochemistry*, Vol. 3. Amsterdam, Elsevier.

Gray, D.J. (2001). Hydrogeochemistry in the Yilgarn Craton. *Geochemistry: Exploration, Environment, Analysis*, 1, 253–264.

Gray, D.J., Noble, R.R.P., and Gill, A.J. (2011). *Field Guide for Mineral Exploration Using Hydrogeochemical Analysis*. CSIRO. Epublish Report No 13936.

Gray, D.J., Noble, R.P., Reid, N., Sutton, G.J., Pirlo, M (2016). Regional scale hydrogeochemical mapping of the northern Yilgarn Craton, Western Australia: a new technology for exploration in arid Australia. *Geochemistry: Exploration, Environment, Analysis*, 16, 100–115. doi:10.1144/geochem2014-333

Gray, D.J., Wildman, J.E., and Longman, G.D. (1999). Selective and partial extraction analyses of transported overburden for gold exploration in the Yilgarn Craton, Western Australia. *Journal of Geochemical Exploration*, 67, 51–66.

Grunsky, E.C. (2010). The interpretation of geochemical survey data. *Geochemistry: Exploration, Environment, Analysis*, 10, 27–74.

Hageman, S.G., Lisitsin, V.A., and Huston, D.L. (2016). Mineral system analysis: Quo vadis. *Ore Geology Review*, 76, 504–522.

Handley, G.A. and Cary, R. (1990). Big Bell Gold deposit. In *Geology of the Mineral Deposits of Australia & Papua New Guinea*, Volume 1 (Ed. F.E. Hughes). The AusIMM, Melbourne, Mono 14, 211–216.

Hillis, R.R., Giles, D., van der Wielen, S., Baensch, S., Cleverley, J.S., Fabris, A., Halley, S., Harris, B., Hill, S.M., Kanck, P.A., Kepic, A., Soe, A., Stewart, G and Uvarova, Y. (2014). Coiled Tubing Drilling and Real-Time Sensing—Enabling Prospecting Drilling in the 21st Century?. *In Building Exploration Capability for the 21st Century Society of Economic Geologists*, Special Publication No. 18, 234–260.

Hoschke, T. (2011). Geophysical signatures of copper-gold porphyry and epithermal gold deposits, and implications for exploration. In *ARC Centre for Excellence in Ore Deposits*, University of Tasmania, 46.

Hallberg, J.A. (1984). A geochemical aid to igneous rock type identification In deeply weathered terrain. *Journal of Geochemical Exploration*, 20, 1–8.

Large, R.R., Gemmell, J.B., and Paulick, H. (2001). The alteration box plot. *Economic Geology*, 96, 957–971.

Leybourne, M.I., and Cameron, E.M. (2010). Groundwater in geochemical exploration. Geochemistry. *Exploration, Environment, Analysis*, 10, 99–118.

Lintern, M.J. (2005). Challenger gold prospect, Gawler Craton. In *Regolith Expressions of Ore Systems* (Compliers:C.R.M. Butt, M. Cornelius, K.M. Scott, and I. Robertson). http://crcleme.org.au/Pubs/Monographs/RegExpOre.html.

Marjoribanks, R. (2010). *Geological Methods in Mineral Exploration and Mining*, (2nd Edition), Springer, 240.

Mazzucchelli, R.H. (1994). Drainage geochemistry in arid regions. In *Drainage Geochemistry, Handbook of Exploration Geochemistry*, Volume 6 (Eds. M. Hale and A.J. Plant), Elsevier, 379–414.

McClenaghnan, M.B. and Paulen, R.C. (2017). Application of till mineralogy and geochemistry to mineral exploration. In *Past Glacial Environments* (Eds. J. Menzies and J.J.M. van der Meer), Elsevier, 689–750.

McQueen, K. (2005). Ore deposit types and their expressions. In *Regolith Expressions of Ore Systems* (Compliers:C.R.M. Butt, M. Cornelius, K.M. Scott, and I. Robertson). http://crcleme.org.au/Pubs/Monographs/RegExpOre.html.

Noble, R.R.P., Lau, I.C., Anand, R.R., and Pinchand, G.T. (2018). *Multi-Scaled Near Surface Exploration Using Ultrafine Soils*. MRIWA Project No. M462. 82p.

Pirajno, F. (2010). Hydrothermal processes and wall rock alteration (Chapter 2) In *Hydrothermal Processes and Mineral Systems*, Springer, 73–164.

Robertson, I.D.M. (1996). Ferruginous lag geochemistry on the Yilgarn Craton of Western Australia: Practical aspects and limitations. *Journal of Geochemical Exploration*, 57, 139–151.

Robertson, I.D.M. (2001). Geochemical exploration around the Harmony gold deposit, Peak Hill, WA. *Geochemistry: Exploration, Environment, Analysis*, 1, 277–288.

Robertson, I.D.M. (2005). Little Eva Cu-Au deposit, Quamby District, Queensland. In *Regolith Expression of Australian Ore Systems* (Eds. C.R.M. Butt, I.D.M. Robertson, K.M. Scott, and M. Cornelius). CRC LEME Monograph, CSIRO, 371–373.

Robertson, I.D.M. and Carver, R.N. (2005). Beasley Creek gold deposit, Laverton district, Western Australia. In *Regolith Expressions of Ore Systems* (Compliers:C.R.M. Butt, M. Cornelius, K.M. Scott, and I. Robertson). http://crcleme.org.au/Pubs/Monographs/RegExpOre.html.

Sabins, F. (1999). Remote sensing for mineral exploration. *Ore Geology Review*, 14, 157–183.

Schodde, R.C. (2014). The global shift to undercover exploration: How fast? How effective? In *Keynote paper for the Society of Economic Geologists 2014 Conference, Keystone Colorado*. www.minexconsulting.com/publications/sep2014b.html.

Schodlok, M.C., et al. (2016). HyLogger-3, a visible to shortwave and thermal infrared reflectance spectrometer system for drill core logging: Functional description. *Australian Journal of Earth Sciences*, 63(8), 929–940.

Sillitoe, R. (2010). Grassroots exploration: Between and major rock and junior hard place. *Society of Economic Geologists Newsletter*, 83.

Smith, R.E., and Pedrix, R.L. (1983). Pisolitic laterite geochemistry in the Golden Grove massive sulphide district, Western Australia. *Journal of Geochemical Exploration*, 18, 131–164.

Taylor, G.F. and Thornber, M.R. (1992). Gossan and ironstone surveys. In *Regolith Exploration Geochemistry in Tropical and Subtropical Terrains*. Handbook of Exploration Geochemistry 4 (Eds. C.R.M. Butt and H. Zeegers), Elsevier, Amsterdam, 139–202.

Whitbread, M. (2002). Ratio analysis of bulk geochemical data: Tracking ore-related cryptic alteration by modelling mineral changes. In *Regolith and Landscapes in Eastern Australia* (Ed. I.C. Roach). CRC LEME Regolith Conference, 133–135. http://crcleme.org.au/Pubs/Monographs/regolith2002/Whitbread.pdf

Witherly, K. (2014). Geophysical expressions of ore systems: Our current understanding. In *Building Exploration Capability for the 21st Century*, Society of Economic Geologists, Special Publication No. 18, 177–208.

3 Sampling of mineral deposits

1 Introduction

The effectiveness and efficiency of all mining operations, and therefore the associated risks, are dependent on correct and meaningful sampling of ore-bodies and surrounding rock. Sampling is important across the entire mining cycle, from exploration, production (grade control), processing (metallurgy) through to environmental purposes, and there are significant costs to the bottom line of a company associated with poor sampling (Carrasco et al., 2004; Dominy, 2016). Although no national or international standard on sampling exists, the Joint Ore Reserves Committee (JORC) Code and other international equivalents do require reporting of sampling issues and ongoing Quality Assurance/Quality Control (QA/QC). Accordingly, there is a need for all technical personnel engaged in the mining chain to understand the science behind sampling and the correct selection of sampling equipment and its use, so as to minimise the errors resulting from it. It is also imperative for mining and exploration company management to appreciate the importance and contribution of correct (and incorrect) sampling to the improvement of the overall cost and intangible benefits to the mining operations of the organisation.

Most operations of a mining cycle involve taking particulate samples of ore or waste materials with the premise that the sample extracted is representative of the larger volume of material it is extracted from (Gy, 1982; Pitard, 1993; Holmes, 2010). Across all stages of the mining cycle, sub-samples are extracted from larger volumes to eventually measure a parameter (e.g. concentration of metals, density of material, etc.). The sampling process reduces the mass but needs to accurately and precisely represent the larger target material or *lot*. During the drilling phase of exploration, sub-samples from every metre or composite metres of Air Core (AC), Reverse Circulation (RC) or other drill cuttings are obtained to eventually provide an assay that allows an estimation to be made for further studies.

During resource development drilling, RC cuttings that represent each metre amount to approximately between 15 and 25 kg, from which a sub-sample is ultimately extracted for final analytical measurement to obtain an assay. Alternatively, full or half diamond core of variable recovery may be sub-sampled for analytical assay, density measurement or metallurgical testing. During the mining production stage (grade control), samples may be taken from blasthole piles in open pit mines, from chip samples across an exposed underground face, rom blasted rock on the surface or from a drawpoint underground. During the storage of ore or blending of ore types, samples are extracted from stockpiles. In the processing plant, samples are obtained off conveyor belts or from slurries and tailings, and waste rock samples are taken for environmental monitoring.

Unfortunately, in all sub-sampling operations within the mining cycle, the smaller sample extracted for the final measurement may not always be representative of the entire population of the target material, and the entire process is subject to errors related to the inherent geological variation of the samples (heterogeneity), and also to the physical operation of extracting, handling and analysing samples.

It is important for all technical personnel involved in sampling across the mining chain to have a basic understanding of the science behind sampling. For example, exploration and mining operations have an established sampling protocol for determining the concentration of the required metals so as to discriminate ore from waste (Figure 3.1). For a grade control sampling protocol, a sample is extracted from a blasthole pile, RC drill rig or chip sampled across an underground mine face (Figure 3.1). The sample is then either split at site, i.e. the mass of the sample is reduced using a splitter, or sent in total to the laboratory, where the particle size is reduced using a variety of crushers and pulverisers in sequence, with the mass of sample being reduced using variety of splitters (riffle, cone, rotary splitters). Finally, the measurement of the analyte of interest is carried out via the analytical step of the sampling chain. During the entire process, smaller samples are extracted from a larger sample via a single increment or multiple increments. The process of reducing the mass of the target material introduces *errors* due to the inherent variability of the natural material, the physical or robotic interaction of equipment to extract a sub-sample, or due to the final analytical measurement (Gy, 1982; Pitard, 1993; Richardson et al., 2005; Minnitt et al., 2007a) (Figure 3.1).

Sampling errors are not self-cancelling and the sampling error variances all add up to produce a total sampling error (Figure 3.1). It is necessary to minimise the errors during the entire sampling process, which is achieved by an understanding and application of Gy's sampling theory, often referred to as the Theory of Sampling (TOS). The TOS recommends a scientific and

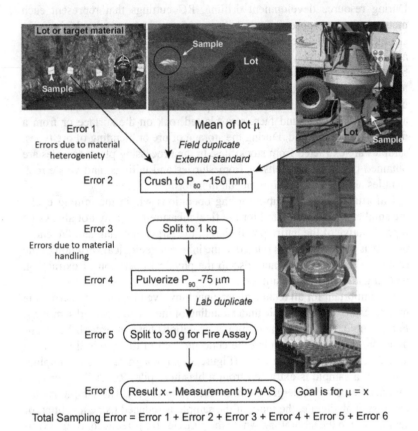

Total Sampling Error = Error 1 + Error 2 + Error 3 + Error 4 + Error 5 + Error 6

Figure 3.1 Concept of sampling errors in a common sampling protocol for grade control during mining. A sample of much smaller mass is extracted as a single or multiple increment from the lot which has a much larger mass (e.g. underground face or blasthole or reverse circulation drill hole), and a sequence of sample handling operations are followed to arrive at final analytical measurement. Most errors arise due to the Earth materials' heterogeneity and the magnitude of the error depends on the sample size taken from the lot. Another set of errors arise due to material handling i.e. nature of extraction at each handling stage of the protocol. The final error is due to analytical measurement practices. All errors add up to the total sampling error.

structured approach to sampling with the assurance that sampling errors are minimised.

2 Theory of Sampling (TOS)

The TOS was first formulated by a French engineer, Pierre Gy (Gy, 1982), and articulated for the resources industry by Francis Pitard (Pitard, 1993) and Dominique François-Bongarcon (François-Bongarcon and Gy, 2002). The theory identifies that error is present because of a difference in the assayed grade value and the true value. The error is due to a combination of factors, including the heterogeneity present in the materials, the manner in which the materials are sampled and handled, including time-dependent sampling practices, and the final measurement.

The *sample* (the sub-sample) taken from the *target material* or *lot* should be representative of the lot. That is, each of the particles in the sample should have an equal *probability* of being extracted from the particles in the lot. This is the concept of probabilistic sampling.

2.1 Types of sampling errors

All errors arising throughout the entire sampling procedure, including the final analytical result, are labelled Global Estimation Error (GSE), which itself is composed of two main errors: Total Analytical Error (TAE) and Total Sampling Error (TSE) (Figure 3.2, Table 3.1). Of the two main error types, the analytical error is due to practices arising from the analytical process or final measurement, and is considered minimal (Minnitt et al., 2007a) in comparison to total sampling error due to strict laboratory controls and the high precision of modern analytical instruments (Wagner and Esbensen, 2015). Therefore, most of the errors in the sampling process are due to the TSE, which in turn is composed of several error components (Figure 3.2, Table 3.1). The error components can be separated into those occurring due to *material heterogeneity* (fundamental sampling error, grouping and segregation error and nugget effect, although nugget effect is not universally acknowledged), errors due to *process variations* (long-range periodic and random fluctuations) and those due to the *sampling processes* dealing with tools and techniques (increment, delimitation, extraction, preparation and weighing errors). The error causes, components and definitions are provided in Table 3.1.

All of the error components can be grouped under two types of errors: *correct sampling errors* (CSE) and *incorrect sampling errors* (ISE) (Figure 3.2, Table 3.1). The correct sampling errors arise due to material properties: constitutional and distributional heterogeneity. *Constituent* (or *compositional*) *heterogeneity* (CH) arises due to the variations in composition,

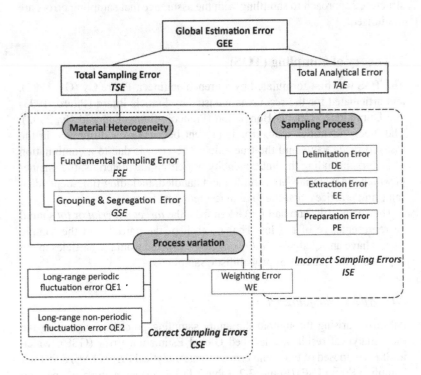

Figure 3.2 Overview of the sampling errors that exist in a stationary lot. Modified
from Richardson et al. (2005). Many of the errors under incorrect sampling errors use the prefix *increment* (e.g. increment delimitation error)
to signify that the error is due to taking a partial sample unit.

shape, size and density of the particles or fragments in the lot. *Distributional
heterogeneity* (DH) is the inherent nature of variation in spatial and physical
distribution of particles in the lot, and results in the differences of measured
parameters (e.g. grade) between groups of particulate materials in the lot.
Broken rock particles have a high tendency to cluster and segregate across
a range of spatial scales within the target material (the lot), and the grouping of particles underpins the DH. For example, denser sulphide-bearing
particles in a fragmented ore are likely to segregate to the bottom of the
broken particle pile, thereby increasing the DH of the target material. Both
these properties arise due to the materials themselves, and although they can
be minimised, they can never be eliminated (contrary to their label). The
incorrect sampling errors are due to the sampling process itself, and include
equipment type and materials handling, and can be reduced to negligible
or eliminated through the proper extraction procedures and equipment use.

Table 3.1 Definition of sampling errors as defined in TOS (from Dominy, 2018, compiled from Gy, 1982; Pitard, 1993). Not listed are the point selection errors arising due to plant processes and longer-term fluctuations.

Sampling error	Acronym	Error type	Effect on sampling	Source of error	Error definition
Fundamental	FSE	Correct sampling error	Random errors	Characteristic of the ore type, due to constituent and distribution heterogeneity	Results from grade heterogeneity of the broken lot. Of all sampling errors, the FSE does not cancel out and remains even after a sampling operation is perfect.
Grouping & segregation	GSE				Relates to the error due to the combination of grouping and segregation of rock fragments in the lot. Once the rock is broken, there will be segregation of particles at any scale (e.g. surface stockpile or laboratory pulp).
Delimitation	DE	Incorrect sampling error	Systematic errors – generate bias	Sampling equipment and materials handling	Results from an incorrect shape of the volume delimiting a sample.
Extraction	EE				Results from the incorrect extraction of a sample. Extraction is only correct when all fragments within the delimited volume are taken into the sample.
Weighing	WE				Relates to collecting samples that are of a comparable support. Samples should represent a consistent mass per unit.
Preparation	PE				Refers to issues during sample transport and storage, preparation (contamination and/or losses), and intentional (sabotage and salting) and unintentional (careless actions and non-adherence of protocols) human error.
Analytical	AE			Analytical process	Relates to all errors during the assay and analytical process, including issues related to rock matrix effects, human error and analytical machine maintenance and calibration.

2.2   *Errors due to material heterogeneity – the fundamental sampling error*

The errors arising due to material heterogeneity are the fundamental sampling error (FSE) and the grouping and segregation error (GSE) (Figure 3.1). *Material heterogeneity* is the main cause of sampling errors. Earth materials are rarely homogeneous, and heterogeneity is the norm. A sample in the mining cycle, such as from a blasthole pile or broken ore from an underground face, when observed from a distance may appear to be homogeneous, but at a microscopic level (or even molecular scale), the Earth material is made up of different particle sizes and compositions. That is, it is heterogeneous in grain or particle size, particle shape, particle mass or density, particle chemical composition and even particle size distribution. The heterogeneity occurs due to differences in composition of the particles, clustering of similar fragments or particles (called grouping) and segregation of particles due to gravity, shape or density.

The CH and DH in Earth materials give rise to the FSE and GSE.

The FSE and GSE are impossible to eliminate due to the inherent existence of CH and DH of the target material, but both errors can be reduced to acceptable levels, by following and correctly implementing proper sampling protocols. The FSE can be minimised by reducing the error based on the target material properties, mainly the grain size of the material and/ or increasing the extracted sample mass, so that more particles of the lot are extracted. The GSE, which is mainly dependent on the segregation of broken particles, can be reduced by thorough mixing and blending the material of the lot prior to sub-sampling, which is often difficult or impossible in practice if large lots need to be sampled, such as stockpiles or blasthole piles. The practical approach is often to extract a number of smaller sub-samples from the lot and combine them (referred to as compositing) to make up a larger sub-sample. By taking a larger number of smaller samples, a random approach to sampling the lot is achieved, thereby reducing (but never eliminating) the GSE.

2.2.1 *FSE equation and parameters*

The FSE of extracting a sample from the lot can be theoretically estimated and modelled if certain parameters of the lot are determined (Pitard, 1993). The FSE equation is:

$$\mathrm{FSE}(\sigma^2) = f.g.c.l.\mathrm{d_N}^3 \left(1/\mathrm{M_S} - 1/\mathrm{M_L}\right) \tag{1}$$

Where

σ^2 is the variance of the sampling error.

f is the *shape factor*, which relates to the volume and diameter of the particles of the material to each other, and depending on the main shape of the particles either has value of 1 (perfect cubes) to 0.1 (needle shapes), but for most practical purposes takes a value of 0.5 (spheres).

g is the *granulometric factor* or size distribution factor, which is the average fragment volume (V) divided by the nominal fragment volume (V_n), which for most purposes is shown to achieve a value of 0.25 for most Earth material types.

d_N is the nominal size of the fragments in the lot or target material. In practice, its value is the screen size that retains 5% of the material in the lot, with 95% of the material passing through the screen. Sometimes the nominal size is referred to as *top size* of material and denoted as a subscript (i.e. d_{95}), which implies that 95% of particles passed through a screen of the noted particle size.

c is the mineralogical factor, and is made up of two components: the ratio of the metal/waste densities and grade of the target material (lot), via the equation:

$$c = \frac{(1-a_L)^2}{a_L} \rho_c + \rho_m (1-a_L) \qquad (2)$$

where a_L is average grade or composition of the analyte (metal) in the lot, ρ_c is the density of the critical particles (ore or mineral particles) and ρ_m is the density of the matrix or gangue component. For low concentration ores such as gold, the approximation is used where

$$c \approx \rho_c / a_L \qquad (3)$$

where ρ_c is the density of the precious metal and a_L is the average grade of the sample. The nominal density of gold is 19.3 g/cm^3.

M_s is the *sample mass* (sub-sample) and M_L is the mass of the lot.

l is referred to as *liberation factor*, and depends upon the size of particles and the degree of liberation of free grains of the metal or analyte under consideration. It varies between 0 for no liberation to 1 for fully liberated metal grains. The factor was originally denoted as $l = (d_L/d_N)^{0.5}$, where d_L is the liberation size of the metal or analyte under consideration (effectively the metal particle size) and d_N is the nominal grain size of fragments. However, the value of the exponent has been debated, especially for gold, where using 0.5 is noted not to provide realistic values (François-Bongarcon and Gy,

2002). The recommendation is to use b as the exponent, with $b = (3 - \alpha)$, where α is determined experimentally from Duplicate Sampling Analysis (DSA). Sometimes a default value of 1.5 is applied for gold ores; however, recent work has noted that the value for many gold ores is closer to unity ~1 (Minnitt, 2017). For non-gold or precious ores, the original equation of l can be used.

Using the liberation factor component, the relative variance error equation for precious metals can be written as

$$\text{FSE}(\sigma^2) = f.g.c. \ d_1^{3-\alpha} \ d_N^{\alpha} \ (1/M_s - 1/M_L) \qquad (4)$$

For most theoretical purposes, M_s is negligible compared to M_L, and taking a constant K for a given ore type to represent $K = f.g.c. \ d_1^{3-\alpha}$, the simplified FSE equation is

$$\text{FSE} \ (\sigma^2) = K.d_N^{\alpha}/M_S \qquad (5)$$

If the exponent for d_N is 3 rather than α, then the constant is termed C.

The sampling constant, K and parameter a can be experimentally determined using three main methods:

1 A sampling tree experiment (FTE) or Duplicate Sample Analysis (DSA) (François-Bongarcon and Gy, 2002; Minnitt et al., 2007b).
2 A heterogeneity test (HT), which is the most commonly one used (Pitard, 1993) and provides only a value for K.
3 Segregation free analysis (SFA) (Minnitt, 2017).

The main relationships denoted by the FSE equation (Equation 4) are

1 The error variance is directly proportional to the cube of the particle size of the lot;
2 The error variance is inversely proportional to the sample mass, given a constant grade or concentration;
3 For samples of different grades, those with higher grades have much lower errors.

Therefore, to reduce the FSE, the options available are either to extract a larger mass from the lot or reduce the particle size prior to extracting a known mass, for a given concentration of the metal.

The FSE equation allows the theoretical estimation of specific parameters prior to sampling, provided that the constant C can be estimated based on testing of the material or K is determined for a specific ore type. Uses of the FSE once C or K are known include:

- Selection of the appropriate mass of the sample to be extracted from the target material or lot.
- Calculation of the the FSE given the particle size and sample mass to be extracted from the lot.
- The degree of particle size reduction required via grinding and pulverising prior to extraction of the sample of given mass within a set FSE.

Examples of the use of a calculated FSE are given next.

Example 1. Blasthole sampling in an open pit gold mine that is hosted by a quartz-rich metasediment host provides the following data:

d_N = 1.27 cm (the particle top grain size is determined using size testing methods, where 95% of the material passes through a particular mesh or sieve size); Au grade = 1 g/t; f = 0.5; g =0.25; liberation size d_l = 10 μm = 10^{-3} cm, and α = 1.5. Given that the density of gold is 19.3 g/cm³, determine the sample mass required to achieve a standard deviation of 10% or better, i.e. FSE (σ^2) = 0.01 and $\sqrt{FSE}$ = 0.1 (10%).

Solution: Re-arranging the FSE equation (Equation 3), and considering that the mass of the lot – M_L – is much larger than that of the sample, the sample size required to achieve a standard deviation of 10% or better is given by:

$$M_S = f.g.c.\ d_l^{3-\alpha}\ d_N^{\alpha}/\ FSE(\sigma^2)$$

Given the case of precious metal – Au, the value of c can be approximated as $c \approx r_c/\ a_L$ which in the example becomes 19.3/1. Substituting the values in the FSE,

$$M_s = \left(\frac{0.5 \times 0.25 \times 10^6 \times 0.001^{1.5} \times 1.27^{3-1.5}}{.01} \right)$$
$$= \frac{10.88 \times 10}{0.01} \text{grams}$$
$$\approx 11 \text{ kg}$$

Example 2. The sample collected from Example 1 i.e. 11 kg, is transported to the lab for a final gold assay so that ore or waste can be defined. The lab has to sub-sample a maximum of 1 kg for further splitting. Therefore, what top grain size should the sample be crushed and ground to achieve a standard deviation of 10% or better i.e. FSE (σ^2) =0.01?

Solution: Alpha (α) for the liberation calculation is taken as 1.5. Rearranging the FSE Equation (3) to determine the top particle size, results in

$$d_N{}^\alpha = M_S.FSE(\sigma^2)/f.g.c.d_1{}^{3-\alpha}$$

$$d_N{}^{1.5} = \left(\frac{.01 \times 1000}{0.5 \times 0.25 \times 0.001^{1.5} \times 19.3 \times 10^6}\right)$$

$$d_N = \left(\frac{.01 \times 1000}{50.07}\right)^{1/1.5}$$

$$\approx 0.13 \text{ cm}$$

Example 3. A geologist in charge of a RC drill programme to assess a porphyry copper deposit needs to estimate the minimal sample mass for each 2 m composite sample. The main copper ore is found to be chalcopyrite ($CuFeS_2$) in the granite host rock. Preliminary metallurgical tests have shown that the chalcopyrite grains have an average diameter of 0.05 cm and that the average copper grade is 1.2%. The RC rig used is shown to produce particles with a top size of 1.27 cm. The density of chalcopyrite was estimated at 4.2 g/cm³ and that of granite as 2.8 g/cm³. What is the minimum weight of the sample required over each 2 m interval to ensure that the variance of the sampling error will not exceed 5% (i.e. 0.05)?

Solution: Re-arranging the FSE equation (Equation 1), and considering the mass of the lot – M_L – is much larger than the sample, the sample size required to achieve a variance of 5% or better is given by:

$$M_S = f.g.c.l. \ d_N{}^3/ FSE(\sigma^2)$$

The known data provides the following in the FSE equation:

d_N = maximum particle size = 1.27 cm
Estimated ore grade = 1.2% Cu (0.012)
Required precision or error $FSE(\sigma^2)$ = 5% (.05)
Grain size of chalcopyrite d_1 i.e. liberation size = .05 cm
Density of chalcopyrite ρ_c = 4.2 g/cm³
Density of matrix (granite) = 2.8 g/cm³
Using standard values for shape factor f = 0.5 and granulometric factor
  g = 0.25.

First calculate the proportion of copper in chalcopyrite ($CuFeS_2$) which is the molecular weight of Cu/molecular weight of chalcopyrite = 63.55/183.52 = 0.346. Therefore, for a grade of 1.2% Cu, the copper grade proportion present in chalcopyrite is 0.012/0.346 = 0.0347 =3.47%. The value of a_L is .034, which allows us to calculate c from Equation 2.

$$c = 4.2 \times \frac{(1-.034)}{.034} + 2.8 \times (1-.034) = 115.6$$

To calculate the liberation factor l, the equation $(d_l/d_N)^{.5} = (.05/1.27)^{.5} = 0.19$ is used.

Substituting c into the main FSE equation, together with the other values, gives

$$M_S = 0.5 \times .25 \times 115.6 \times .19 \times (1.27)^3/.05$$
$$= 2350 \text{ g}$$
$$\approx 2.35 \text{ kg}$$

2.2.2 Sampling nomogram

Most AC sampling and assay operations during production have a sampling protocol that can be represented as a sample process flowsheet (Figure 3.1). The optimised sampling protocol should provide a sample mass reduction and sample particle size reduction strategy that conforms to the FSE as calculated using the TOS. The sampling protocol should be such that the total error stays within predefined FSE error limits. Such a strategy can be constructed using the FSE equation as shown in Examples 1 and 2 to calculate particle sizes and sample mass for each stage of the sampling protocol, in order to stay within the predefined FSE error, with consideration that the total sampling error is cumulative. The graphical representation of the sampling strategy that considers the sample weight (M_S) and the nominal particle size or top size (d_N) is referred to as the sampling nomogram, and is generally plotted on a logarithmic scale (Figure 3.3).

The nomogram generally has the predetermined level of FSE error variance displayed, and sometimes multiple levels of FSE error e.g. 10%, 20%. The acceptable error line is known as Gy's safety line. Sampling procedures in the sampling chain that fall above the safety line are considered to have higher error than accepted and are clearly identified from the nomogram. The example of the nomogram in Figure 3.3 shows that a sample split via rotary splitter from the blasthole sample, with a total mass of 90 kg (the lot) having a nominal particulate grain size of 3 cm has a very high error and the subsequent split (Split 2 in Figure 3.3) also has a high error. Only after the Crush 2 stage, where the particle size has been reduced to 0.3 cm, does the error variance fall within the acceptable threshold. The nomogram allows the geologists or laboratory staff to take remedial action at particular

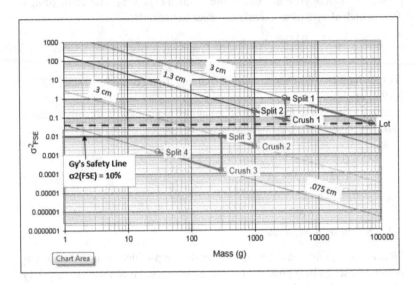

Figure 3.3 Sampling nomogram. The nomogram shows main sample mass reduction and grain size reduction steps with the corresponding fundamental sampling error variance (σ^2_{FSE}). The safety line in this example is set at 10% (.01) σ^2_{FSE} and another at 14% (.02, dashed line), with individual splitting steps (mass reduction) and grain size reduction steps (crush) at 3 cm, 1.3 cm .3 cm and .075 cm. The nomogram shows that split 1 will result in a large error, whereas splits 3 and 4 are within the predefined error given that the incorrect sampling procedures during splitting and grinding are eliminated.

sampling stages to improve the sampling process and minimise the correct sampling errors, which in the example are at split stages 1 and 2.

2.3 Errors due to material handling – incorrect sampling errors

Incorrect sampling errors arise due to equipment interactions with the target material and the extracted sample, and due to poor preparation procedures (Figure 3.2). Unlike the correct sampling errors, these errors are largely avoidable, and can be greatly reduced and even eliminated by setting of a proper sampling protocol, but importantly, practically implementing that protocol. Sampling protocols require that a sub-sample is taken from the lot and often at various later stages in the sampling chain; for example, sampling a blasthole pile followed by crushing followed by sub-sampling to reduce the mass. The process of extraction of the sub-sample or the increment is the cause of several of the ISE, as noted in Figure 3.1 and Table 3.1.

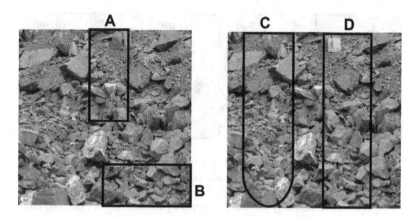

Figure 3.4 Concept of delimitation error. The images show broken ore with partially segregated particles (smaller at top, larger at bottom). The increments selected using the sampling devices in A and B do not representatively sample the segregated particles with A preferentially sampling finer particles and B sampling coarser ones. In sampling scenario C, using a round bottom sampling device incorrectly samples the segregated particles, biasing more to the lighter and finer particles segregated to the top of the pile. Increments selected with a rectangular shaped device through the entire broken pile correctly extract the sample by equally selecting segregated particles.

Source: Concept from Pitard (1993) and Gerlach and Nocerino (2003).

Delimitation error (DE) occurs when the sampling device or operation fails to provide an equal probability for all parts of the lot to be sampled. The sampling device design and operation must provide an equal and constant chance for all of the particles in the lot to be extracted. An example of DE due to device design and operation is provided by blasthole piles, where the correct round bottom trays do not provide an equal chance of selecting all particles in a pile, whereas the use of a rectangular tray or scoop allows a much higher probability of selecting representative sample of the pile (Figure 3.4).

Extraction error (EE) occurs due to the interaction of the sampling tool with the particles in the lot. The sampling tool defines the boundary between the sub-sample and the lot, whereby all particles having their centre of gravity inside the tool boundary are supposed to be part of the sub-sample and all particles outside the tool boundary are retained within the lot. The sampling tool during operation cuts a path that is unable to collect or divide particles in its path and therefore particles that were supposed to be extracted are left behind (Figure 3.5). To reduce the extraction error, a correctly designed

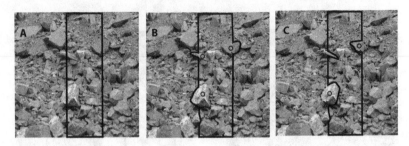

Figure 3.5 Concept of extraction error. A. The correct extraction or target boundary that cuts across some particles, especially coarser ones. B. Ideal extraction where coarser particles having their centre of gravity within the device boundary are sampled (circled particles). C. An extraction error occurs where the particles that lie along the boundary of the sampling device are not extracted. The particles denoted by small circles whose centre of gravity lies within the sampling device boundary should have been extracted but have not been so. Concept from Pitard (1993).

sampling device should be used and operated, whereby the device should penetrate the lot at a slow, even rate.

Preparation errors (PE) arise due to a range of issues with sample handling, sample integrity and preservation, along with human error. The PE can occur due to:

- Contamination of the samples from other particles outside of the sample or from the next (previous) sample.
- Samples suffer losses due to poor storage equipment, such as adhesion to the sides of bags or losses from bags as fine particles.
- Samples suffer extreme unintended physical and chemical alterations due to crushing, pulverising, sieving and drying, or oxidation of reactive samples.
- Samples are subjected to deliberate alteration such as sabotage in the form of salting and wrong delimitation.
- Samples suffer from human errors such as mixing of labels, dropping of sample and general negligence.

3 Sample types and sampling methods

Various sample types and procedures are used across the entire mining cycle depending on the mining operations (open pits versus various types of underground mining), ore types and scales of operation. Although some sample types described are common across surface (open pit) and underground grade control, such as diamond drilling and grab sampling, there are some sample types, such as channel and panel sampling, which are unique to underground mining operations. In the collection of each sample type, the errors noted in the TOS and common to the sampling process need to be carefully considered and minimised, although the extraction of some sample types, by their very nature, preclude minimising errors.

3.1 Grab sample

Grab samples are a single large sample or a series of smaller samples combined into a larger one from blasted ore piles (muck pile), from rock faces, trucks, from ore stockpiles or from conveyor belts and slurries. The samples are generally collected by hand, shovel or bucket. The amount or mass of a sample is variable and depends on the sampler, time constraints and laboratory capacity, with 1 kg–5 kg samples during grade control being common. Grab samples are mostly used due to access issues or safety concerns with other sampling methods or a lack of other sample data (Dominy, 2010).

Grab samples are considered to manifest a high bias and technical staff collecting such samples need to be aware of the pitfalls so as to minimise FSE, GSE, DE and EE errors, the main issue is that of representation (Dominy, 2010). Firstly, there is the human bias where samplers tend to select samples based on visible indications of high grade, and also preferentially sample fines. Secondly, as grab sampling is suited to collection from muck piles, high- or low-grade material can preferentially segregate into piles, thereby resulting in preferential sampling. Thirdly, muck piles are often zoned during creation, with the last material being present on the surface and so often preferentially sampled.

Grab samples should be avoided, but where they are the only sample available, especially in specific mining methods such as block caving or where safety demands it, the following procedures are recommended.

- Collect a large number of smaller samples (at least 32) to minimise the GSE, and even use larger sample sizes in specific troublesome deposit types, such as coarse gold systems.
- Proper characterisation of the mineralisation is necessary, especially the determination of the relationship between fragment size and grade.
- Undertake a heterogeneity test (HT) so as to determine the accepted sampling precision via the sampling equation.

3.2   *Chip and panel sampling*

Chip samples are individual rock samples collected across a face, either at regular intervals or within geological boundaries. The samples are often composited to one sample. The usual method is to measure and mark the sample location along the face, place a tarpaulin beneath the face being sampled, and break off rectangular bits of rock at the marked points, then collect the entire broken 'chips' as a single sample. Another approach is to break the face into geological zones a few metres across (i.e. hanging wall, ore zone, footwall), and sample each separately (Figure 3.6). The chip sample line direction can be either horizontal or inclined depending on the direction of mineralisation, although the sample line should be normal to the orebody.

A chip sample, if taken properly, is far more representative than a grab sample. The challenge then becomes to ensure that a similar amount of material is broken off the face each time (rectangular volume), otherwise extraction and weighting errors are introduced and increased (Figure 3.6). Often due to variability in operator training and commitment, and rock hardness, there may be a difference between masses collected for each chip sample due to either lesser amounts being broken or excess being broken due to over-chipping (resulting in EE and WE) (Figure 3.6). Operators are also tempted to collect the easily broken material, preferentially resulting in delimitation and extraction errors in the sample. Therefore, during chip sampling, it is important to consider the amount of sample being broken and its true representativeness with respect to the requirements as dictated by mineralisation type.

Panel sampling is an extended version of chip sampling done on a face divided into a rectangular grid (Figure 3.6). The rock face is divided into quarters and marked, and then chips are broken off each quarter to add to the entire sample. The samples generally fall onto the tarpaulin and are combined into one. Panel sampling involves taking more increments of sample than the single line chip sampling, and so has less bias (lower PE), but involves greater effort and a greater chance of EE errors that occur during chip sampling. The extra effort involved in chipping a larger amount of sample can cause the sampler to tire towards the end of a panel, and try to rush through the final bit of the face, thus creating a PE with a bias between the end and the start.

Although there are genuine concerns with the biases introduced during chip and panel sampling, the ease of collection in the confined underground environment is the reason for their common use. If chip sampling is employed, good sampling practice can minimise errors to acceptable limits. Care should be taken to:

• Clean the sampling area and conduct a hazard check.
• Identify the mineralisation and make a face geological map where rock variation allows it, noting the ore-waste boundaries.

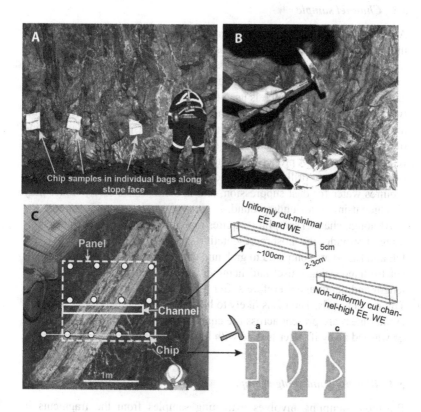

Figure 3.6 Linear sampling. A. Geologist taking chip samples across a stope face
with stope divided according to geological features and sample locations
(with bags) marked. B. The chip sampling process where rock pieces are
chipped with a hammer into a sample bag. C. Chip, panel and channel
sample concept. Sample handling errors, such as extraction error (EE)
and preparation error (PE), are introduced for chip sampling when the
exact amount of sample a is not taken as shown in b and c. For channel
samples, if the volume of sample cut is not uniform across the channel
length, this can result in high EE.

- Measure the section across mineralisation (if the ore is narrower than
  the face) and mark out the chip sample locations, including the length
  support of each chip sample where necessary.
- Conduct chip sampling, making sure an equal volume of rock is collected from each sampling location, preferably with the depth being
  kept constant as allowed by the rock properties.
- Carefully label the sample bags followed by secure delivery to the lab.

3.3   Channel sample

A channel sample requires the operator(s) to cut a solid block of rock of uniform volume across the rock face (Figure 3.6). Channels are cut either by hand or by using a diamond saw. The sample is generally collected by making two parallel cuts approximately 2–3 cm deep and 5–10 cm apart and a hand chisel or air pick is used to break the block of rock between the two cuts. A channel cut by hand is more difficult to keep even, whereas for a saw cut channel it is easier to maintain the dimensions. The section line of the channel on the face should be selected so as to provide the most representative sample. A properly cut channel sample is the closest one can get to a diamond core drill sample without actually drilling. The use of a saw requires water for dust suppression, and in some mines safety issues may preclude using a saw underground.

Although channel sampling requires more effort and time than chip sampling, it provides the most representative sample, with comparatively lower DE and EE errors compared to grab and chip samples. For channel samples cut by hand using chisel and hammer are difficult to maintain constant length and depth and therefore suffer a higher DE and EE as compared to saw-cut channels. The EE is likely to be high to moderate if rocks of different hardness are present across the channel, and softer material is likely to be washed away if water is used with the saw.

3.4   Blasthole and sludge sample

Blasthole sampling involves extracting samples from the fragments in drilled rock piles surrounding a blasthole. The holes are drilled to accommodate explosives for subsequent blasting and therefore are not tailored to grade control sampling, but are invariably used in open pit grade control (for non-precious metal mines) due to their availability i.e. often negligible cost. The sampling of the blasthole piles requires an understanding of the reason for the blasthole. In a typical surface mining operation, the blasthole is drilled to blast a flitch or bench and the spacing and geometry of the holes is dictated by rock mechanics rather than grade or mineralisation. The hole surface is represented by drilled material as a cone (rock cuttings), with most of the hole depth equivalent to the bench height and a sub-drill portion drilled into the underlying bench for purposes of blasting mechanics (Figure 3.7A). The drilling geometry results in vertical and lateral segregation of the drilled cuttings (Figure 3.7A). The top of the blasthole cone is composed of sub-drill material that needs to be avoided, the middle and the majority of the cone is made up of the middle to lower part of the bench material, and the bottom of the cone is composed of material from previous sub-drill, and thus recovery is poor. At the surface, the blasthole drilling systems result in loss of fines and segregation of particles across the surface of the cone. An understanding of the segregation is critical in evaluating the use of blast-holes for geological and assay sampling.

Blastholes are often sampled via a tube or spear, shovel or rectangular tray or sectorial sampler. The advantages of blasthole sampling are:

1   Cost. The blastholes are drilled as part of the blasting programme, and therefore bear no cost for the geology team conducting grade control.
2   Resolution. Blastholes are drilled at predetermined spacings ranging from 5–15 m, and so provide a much better spatial resolution than dedicated grade control sampling.

The disadvantages of blasthole sampling are:

1   Due to segregation of the particles in the pile according to depth, but also according to particle size during drilling, spear and shovel sampling result in non-representative and incomplete sampling, and thereby provide high FSE, GSE, EE and DE errors. Trays and other sectorial samplers are better for representative sampling, but the extraction operation needs to be carefully conducted, with the tray making a complete cross-sectional cut into the pile i.e. taking an entire slice of the blasthole pile cone from the outermost edge to the hole opening (Figure 3.7 B, C, D).
2   The sub-drill portion is drilled for blasting practicalities and forms part of the surface pile, but not part of production for the bench, and therefore the sub-drill portion being part of the surface pile for sampling increases the DE. Furthermore, sub-drill portion of the previous bench often is the top of the bench below and so the sub-drill already has broken material and recovery compared to the lower portion of the bench will be comparatively poor and thereby increase EE.
3   The blastholes are drilled vertically and do not consider the geological parameters of the ore, such as orientation, geometrical and grade variations.
4   The drill density and location are dictated by the blasting logistics and programme not by geological or grade control criteria or the drilling programme for the deposit. Therefore, often turnaround times for blasthole assay programmes can be slow, resulting in the geological team being unable to effectively use blasthole samples to delineate ore from waste.

The difficulties in representatively sampling blastholes, including the potential occurrence of very high errors for DE, EE, WE and PE, have prompted several practitioners to recommend avoiding blasthole sampling for grade control and in turn use RC drilling (Pitard, 2008), although some have argued that besides

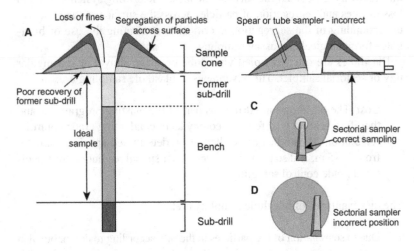

Figure 3.7 Blasthole sampling concepts. A. The blasthole drilling geometry results in vertical and lateral segregation of down-hole materials within the surface blasthole sample cone (modified from Pitard, 2008). B. Section of the blasthole cone that is generally sampled using a spear or tube, which results in un-representative samples with high DE and GSE, whereas the sectorial sampler (tray device) is better. C. Plan view of the blasthole cone showing the correct positioning of the sectorial sampler and D. Plan view of blasthole cone showing incorrect position of sampler to collect sample.

the better geological resolution provided by RC, there is not much difference between the two drilling techniques, and the cost advantage of blasthole is critical (François-Bongarcon, 2010). A review of blasthole sampling versus RC sampling during grade control found that it was unclear whether RC drilling provided a significant advantage in cost over blasthole during grade control, but there was a clear improvement in grade control classification with RC when the ore system was complex, such as gold ore types (Engström, 2017). Therefore, for simpler, less heterogeneous ore types (e.g. bedded iron ore, bauxite), the minimal cost and spatial resolution of blasthole drilling outweigh the potentially higher but controlled errors, whereas in complex ore systems, the greater control of RC in defining the ore-waste boundary and lower DD and EE errors are more critical to the bottom line than lower cost but higher error blasthole.

3.5 *Reverse circulation*

Reverse circulation (RC) drilling is commonly used in resource development and during surface mining, and is now available for underground mine grade control sampling. Compared to diamond core, it provides poorer sample quality i.e. only uniform-sized rock cuttings, but does provide a larger sample mass per metre (approximately 20–40 kg/m) and is faster and cheaper. Newer RC systems come with an attached rotating cone splitter that permits the collection of an unbiased sub-sample, although some still use a static or tiered rifle splitters.

RC drill samples, however, can result in high FSE, DE and EE errors, due to the following:

- They suffer from poor sample recovery and high EE where softer material is encountered. The extraction of clay rich samples or samples below the water table are problematic, as recovery is poor and clogging of cyclones and splitters is common, thereby resulting in high DE and EE errors for some sample types. The loss of sample fines from the collection system is an often unavoidable EE.
- Although a sample is considered to be a composite of a 1 m depth interval, poor drilling practices (visual judgements by drillers of depth) can result in the lot collected not representing a metre and each lot being of variable mass (variable length). Sub-samples will therefore also be of variable mass, when the assumption is that they are from a constant mass lot. The variability of collecting sub-samples along an entire hole will result in different supports to each sub-sample. This results in an increase in DE and WE.
- The extraction of RC sub-sample from the entire metre of cuttings can suffer from a high FSE because a relatively smaller mass is extracted compared to the particle size and overall mass of the lot (1 m).
- Although the attachment of the rotating cone splitter provides relatively unbiased splitting, the levelling of the splitter is important, otherwise preferential sampling occurs, increasing the EE. The tiered or multi-stage splitters are considered biased, with moderate EE as the sub-samples from the same split side are fed into next splitting stage.

For particular ore systems that show variable local geological heterogeneity, especially gold ore systems with narrow steeply mineralised zones (Figure 3.8) and many base metal mines, RC is the preferred choice as it provides much better depth resolution and reduces DE and EE compared to blastholes (Figure 3.8). The advantages of RC drilling over blasthole drilling for grade control sampling are:

- The RC down-hole parameters of dip (angled) and sampling depth and interval can be controlled to provide a far superior definition of the local geology than that achieved from blasthole sampling.

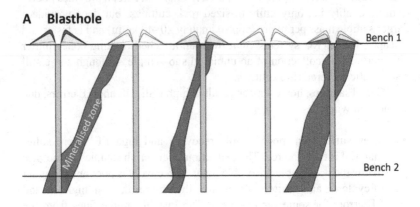

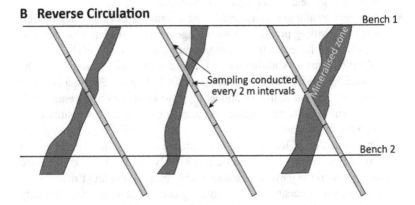

Figure 3.8 Blasthole versus Reverse Circulation (RC) grade control sampling.
A. The blasthole cone sampling for steeply dipping or inclined mineralised ore can be problematic given the bulk sample taken from the
vertical blasthole is unable to define ore geometry and grade with depth
and thereby increase grade errors. B. The RC angled holes with samples taken every specific interval e.g. 2 m, allows the definition of the
inclined ore zone much better, thereby minimising error and optimising
the definitions of ore and waste boundaries.

- The depth control and sample integrity, due to samples returning from
  the inner drill rods in RC drilling, greatly minimises sampling errors
  when compared to blastholes.
- Drillhole layout and scheduling can be optimised for the orebody
  geometry and production deadlines, as opposed to blasthole whose layout and timing are dictated by blasting protocols.

- Although RC drilling for grade control sampling and geology control in gold ore systems is more expensive, studies comparing the two methods suggest that the greater cost of RC drilling is recovered over time due to its superior geological and sampling control to define ore-waste boundaries and minimise sampling errors (Pitard, 2008). However, in relatively less heterogeneous ore systems, there does not appear to be any superiority of RC over blasthole (Engström, 2017).

3.6   Diamond core

Diamond core (DC) drill samples are the best-quality samples because they provide a continuous or near-continuous record of the lithology and mineralisation. The continuous sample produced by DC allows most errors to be generally reduced. Often only half core is extracted as a sample due to QC requirements for further analysis and for preserving the geological record. Diamond core is generally sampled by cutting the core in half using a diamond saw or with a semi-automatic cutting unit. Some of the errors that can arise in DC are:

- Depending on the core diameter, the sample mass can be an issue as smaller core sizes (BQ) provide a limited amount of sample mass (~3 kg/m), and therefore the sample may not be representative of the local mineralisation.
- Core recovery is critical as poor recovery (< 90%) results in an increase in EE.
- Sub-sampling of core or cutting in half in many cases is subjective to the geologist, especially for specific ore types, and introduces bias.

Compared to other sampling methods including RC drilling, DC is expensive and time-consuming.

3.7   Sampling streams of particles

Broken ore, when transferred to the processing plant, is generally passed through the different processing stages as a moving *stream* of dry ore slurry or as a wet ore slurry. Ore as streams needs to be sampled so that the entire stream is cut at regular intervals. Collecting only part of the stream results in bias.

Sampling streams is best achieved via the following principles (Holmes, 2004):

- Sampled at the discharge point of conveyor belt, chute or pipe.
- The complete cross-section of streams must be sampled and this is generally best achieved via cutters.

- The cutter should take a complete cross-section of the stream, moving at right-angles to the stream.
- The cutting time at each sampling point should be the same and the cutter velocity should be the same when it cuts through the stream. The cutter speed should not exceed 0.6 m/s.
- The cutter should be non-restrictive and self-clearing so that each increment sampled is cleared completely and no cross-contamination or weighting occurs between each increment.
- Cutter size and volume should be sufficient to accept the entire increment so as to avoid spillover or loss of sample.
- Manual sampling should be avoided due to safety issues.

Example 5. A shear zone hosted gold deposit resource grade was found to be 4.4 g/t. Much of the resource estimation data was based on 25 × 25 m diamond hole (DH) core sampling (NQ sized core) and finished with traditional fire assay. However, during production, severe variations in gold grades to those estimated for blocks were encountered. What potential sampling issues and QA/QC would need to be checked to find the reasons for the poor reconciliation of estimated versus produced grades?

Solution: The type of deposit and sample collection needs to be scrutinised. Given the shear zone hosted gold, the first investigation should focus on whether the gold is nuggetty (high levels of coarse gold), which can result in severe variations in assays if half core samples are taken i.e. the sampling amount or support is insufficient. If coarse gold is present, then make sure if the larger part of the core can be sampled and the assay type changed to screened fire assay, which is better for nuggetty gold. Secondly, investigate the issues with diamond core as noted in section 3.5. For example, is the core diameter sufficient given the nature of the deposit? Has there been sufficient core recovery? An independent evaluation of the entire sampling process found that the core recovery was very poor given the large amount of shearing in the mineralised zone (recoveries between 50–70%), which led to very high EE, and these were not picked up by inexperienced geologists and core cutting personnel.

Example 6. Grade control sampling in an underground shear-zone-hosted gold was conducting grab sampling by taking 2 random samples from the blasted muck pile of each face. The deposit consistently showed up as 20–25% under-calls during reconciliation between the resource grade and mill grade. What potential sampling issues and QA/QC would need to be checked to find the reasons for the poor reconciliation of estimated versus produced grades?

Solution: The issue is mainly to do with reconciling the grade estimated during resource sampling and plant grade. Therefore, the need is to check the sampling protocol during grade control. Grab sampling for highly

Table 3.2 The sample types – definitions, pros and cons and errors (modified after Dominy et al., 2018)

Sampling method	Definition	Pros and cons	Errors
Broken rock			
Grab	A random pick of rock sample from a broken rock pile	Easy and cheap to collect. Variable mass can be collected. Not representative.	High FSE, GSE, EE and DE
Linear			
Chip	Individual rock chips collected at discontinuous or continuous intervals across a face combined to form a sample	Easy, cheap and fast to collect. Width and depth of sample zone difficult to keep constant.	Moderate DE, high EE and moderate WE
Panel	Multiple chip-channel samples across a panel	Less easy to collect, but more time than chip. Large sample size with better representative. Extraction amounts vary.	Moderate EE, DE and WE
Channel	Constant volume cut across a face using a diamond saw or by hand	Requires more effort, slower. Higher cost, but better at keeping constant depth/weighting.	Low DE and WE and moderate EE
Drilling			
Diamond core	Solid full or half core sample across a specified length	Good geological information. Relatively costly.	Low EE, DE and WE
Reverse Circulation	Cuttings over a specified length but from inner tube (1 m or composited 2 or 4 m)	Larger mass. Cheaper than diamond core. Slightly larger FSE than core as smaller mass extracted from lot. Higher DE and EE due to loss of fines and poor recovery in softer rocks and wet conditions.	Moderate FSE, low EE, DE
Blasthole	Cuttings across specified length (> 2m) with samples from outside of drill rods with sub-drill	Cheap. Provides better spatial resolution but poor depth resolution. Very high errors with poor sample resolution and contamination.	Moderate to high DE and EE
Sludge	Cuttings across depth of face (~1–2 m)		
Stream			
Cutters	A cross-section of a moving stream is intersected at regular intervals	Higher cost, automated with lower errors.	Low EE, DE

heterogeneous ore is prone to acute sampling errors due to representivity problems (see section 3.1). A 2 kg grab sample – most likely biased to larger ore fragment – from the broken ore face (muck pile) is not likely to be representative. Furthermore, gold present in the shear zones may have been concentrated into finer particles after blasting. This problem is solved by shifting to chip sampling of the face before blasting.

Example 7. A lead-zinc stratiform-hosted deposit was being mined by open pit. The deposit was hosted by steeply dipping (~78°) mineralised zones of galena and sphalerite interbedded with shale and siltstones. Grade control was conducted using 15.2 cm diameter blastholes drilled 8 m deep, with 1 m of sub-drill and a spatial pattern of 4 by 4.5 m. Blasthole sampling was conducted via a spear method, where a PVC pipe is inserted into the blast cone taking several increments around the hole (see Figure 3.6). Reconciliation between grade control ore tonnes and grade, resource development DH estimation and net smelter return (NSR) was poor. What potential sampling-related issues should be investigated to improve the reconciliation?

Solution: There are two potential issues, both to do with blastholes. The first is the sampling method for blastholes, where a tube or spear is being used to sample through the blast cone (see section 3.4). The spear is not the best method to sample blast cones and often due to poor practice the sampler does not push the spear through the entire cone. Using the sectorial sampler instead of the spear could solve the problem. The second is to consider employing RC drilling as a complement to the blasthole. RC drilling is recommended because of the near vertical nature (steep dips) of the mineralised zones, whose width and grade variations can be poorly represented by vertical blasthole drilling (section 3.4).

References

Carrasco, P.C., Carrasco, P., and Jara, E. (2004). The economic impact of incorrect sampling and analysis practices in the copper mining industry. *Chemometrics and Intelligent Laboratory Systems*, 74, 209–214.

Dominy, S.C. (2010). Grab sampling for underground grade control. *The Journal of the Southern African Institute of Mining and Metallurgy*, 110, 277–287.

Dominy, S.C. (2016). Importance of good sampling practice throughout the gold mine value chain. *Mining Technology*, 125, 129–141.

Dominy, S.C., Glas, H.J., O'Connor, L., Lam, C.K., Purevgerel, S., and Minnit, R.C.A. (2018). Integrating the theory of sampling into underground grade control strategies. *Minerals*, 8, 232, 45.DOI:10.3390/min8060232

Engström, K. (2017). A comprehensive literature review reflecting fifteen years of debate regarding the representativity of reverse circulation vs blasthole drill sampling. *TOS Forum*, 7, 36–46.

François-Bongarçon, D.M. (2010). Are blasthole samples that bad and reverse circulation samples that good for open pit grade control? Story of a trade-off. In *Proceedings of Sampling and Analysis Conference*, Perth, WA, 23–34.

François-Bongarçon, D.M. and Gy, P.M. (2002). The most common error in applying Gy's formula in the theory of mineral sampling and the history of the Liberation factor. *The Journal of the Southern African Institute of Mining and Metallurgy*, 102, 475–479.

Gerlach, R.W. and Nocerino, J.M. (2003). *Guidance for Obtaining Representative Laboratory Analytical Subsamples from Particulate Laboratory Samples*. EPA Technical Report. Report number: EPA/600/R-03/027. 134p.

Gy, P.M. (1982). *Sampling of Particulate Materials: Theory and Practice* (2nd Edition). Elsevier, Amsterdam, The Netherlands, 431.

Holmes, R.J (2004). Correct sampling and measurement – the foundation of accurate metallurgical accounting. *Chemometrics Intelligent Laboratory Systems*, 74, 71–83.

Holmes, R.J. (2010). Sampling mineral commodities: The good, bad and the ugly. *The Journal of Southern African Institute of Mining and Metallurgy*, 110, 269–276.

Minnitt, R.C.A. (2017). A version of Gy's equation for gold bearing ores. *The Journal of the Southern African Institute of Mining and Metallurgy*, 117, 119–130.

Minnitt, R.C.A., Rice, P.M., and Spangenberg, C. (2007a). Part 1: Understanding the components of the fundamental sampling error: A key to good sampling practice. *The Journal of the Southern African Institute of Mining and Metallurgy*, 107, 505–511.

Minnitt, R.C.A., Rice, P.M., and Spangenbeg, C. (2007b). Part 2: Experimental calibration of sampling parameters K and alpha for Gy's formula by the sampling tree method. *Journal of the Southern African Institute of Mining and Metallurgy*, 107, 513–518.

Pitard, F.F. (1993). *Pierre Gy's Sampling Theory and Sampling Practice*, CRC Press, Boca Raton, FL, USA.

Pitard, F.F. (2008). Blasthole sampling for grade control: The many problems and solutions. In *Sampling 2008 Conference Proceedings, AusIMM*, 15–22.

Richardson, L., Minkkinen, P., and Esbensen, K.H. (2005). Representative sampling for reliable data analysis: Theory of Sampling. *Chemometrics and Intelligent Laboratory Systems*, 77, 261–277.

Wagner, C., and Esbensen, K.H. (2015). Theory of sampling: Four critical success factors before analysis. *Journal of AOAC International*, 98, 275–281.

4   Reporting of mineral resources and reserves in Australia

Introduction

Since the prime asset of any mining company is its mineral deposit(s), it is essential that the extent, nature and quality of each deposit is defined as accurately as possible. While many resource and reserve classification and reporting schemes have been devised, there is a growing interest in and demand for more comprehensive internationally accepted schemes in addition to the existing Committee for Mineral Reserves International Reporting Standards (CRIRSCO) template.

Current resource and reserve regulatory bodies include the Joint Ore Reserves Committee (JORC) Code of the Australasian Institute of Mining and Metallurgy (AuslMM), Australian Institute of Geoscientists (AIG) and Minerals Council of Australia, the CIM Definition Standards on Mineral Resources and Ore Reserves in Canada, the Pan European Reserves and Resources Reporting Committee (PERC) of the Institution of Materials, Mining and Metallurgy (IMMM), European Federation of Geologists (EFG), Institute of Geologists of Ireland (IGI), and the United Nations Economic Commission for Europe (UN-EGE) Framework Classification.

There is also a harmonisation process in place with the objective of developing a truly international code. Since the PERC Code is very largely based on the JORC Code and has improved on it in many areas, it is being suggested as a suitable basis for such an international code.

For many companies around the world, the JORC Code is still seen as the de facto standard, and its use is mandatory for all listed entities on the Australian Securities Exchange. The JORC Code was first released in 1989, with the Guidelines to the Code published in 1990. Both documents were revised and released in a combined form in 1992, with an Appendix dealing with diamond deposits issued in 1993. A further minor revision took place in 1996, and the last major revision was completed in 2012.

The JORC Code establishes minimum standards of reporting mineral resources and ore reserves in Australasia. Its adoption ensures that public reports dealing with such matters are appropriately informative to investors and their advisers. The Code is applicable to all minerals, including metal-liferous ores, gemstones and coal, with its main features being:

- The classification of tonnage (or volume) and grade (or quality) esti-mates of a mineral deposit as mineral resources or ore reserves with appropriate levels of confidence.
- The identification of the so-called Competent Person or Persons endors-ing the estimates and their qualifications and experience.
- The definition of the responsibilities of the Competent Person(s) and Boards of Directors regarding the reporting of mineral resources or ore reserves.

Both the Australian and New Zealand Securities Exchanges have adopted the JORC Code in full as part of their listing rules for minerals companies. The Australian Securities Exchange Limited (ASX) listing rules also define a regime for continuous disclosure and periodic reporting for all listed com-panies. Failure to do so can expose the company and its Directors to liability under the Corporations Law.

The guiding principles of the JORC Code are transparency, materiality and competence. **Transparency** requires clear, explicit and unambiguous reporting with concise but sufficient information provided. The reporting must not be misleading. **Materiality** requires technical information to be of appropriate detail and relevance to investors, with a balanced use of historic information and estimates. **Competence** means that public reporting of Min-eral Resources and Ore Reserves must be compiled by a Competent Person.

A Competent Person is a Member or Fellow of the Australasian Insti-tute of Mining and Metallurgy (AusIMM) and/or the Australian Institute of Geoscientists with a minimum of five years' experience relevant to the style of mineralisation and type of deposit under consideration and the activity undertaken. The ASX Listing Rules stipulate that the Competent Person must be identified along with his/her employer and the report must also include a statement to the effect that the information on mineral resources and/or ore reserves was released with the permission of the Competent Person(s).

The main reasons for the success of the JORC Code are:

- Its regulatory authority and support by the main securities exchanges.
- Its relative simplicity and avoidance of complex definitions and opera-tional detail.
- The industry's willingness to audit the system and discipline the Com-petent Person(s) when necessary (Stephenson, 2000).

The ASX's adoption of the JORC Code gave it national credence and regulatory authority, whilst the freedom given to the Competent Person(s) to apply their experience, skills and professional judgement makes it user friendly. Competent Person(s) are held accountable for their estimation of mineral resources or ore reserves and actions by the relevant professional bodies (i.e. AuslMM or AIG). Both have Ethics Committees responsible for policing the professional actions of their members.

Apart from the responsibilities of the competent person(s) in estimating mineral resources or ore reserves (Phillips, 2000), company directors and mining professionals have a responsibility to disclose relevant information about changes in these estimates which could have an effect on the price or value of the mineral asset to the ASX (Listing Rule 3.1). Failure by a listed company to do this could expose it and its directors to liability under the Corporations Law. ASX listing Rules Chapters 4 and 5 also require periodic disclosure and emphasise the need to report on mining and exploration activities, with reports on resources and reserves being required by Appendix 5A of the Listing Rules, which contains the JORC Code.

The JORC and other codes

The JORC Code is the Australasian Code for Reporting Mineral Resources and Ore Reserves. It is well accepted internationally and is used as the benchmark for both internal and external reporting by some international mining companies. It was also the basis of agreement by Australia, Canada, South Africa, United Kingdom and United States, through the Council of Mining and Metallurgical Institutions (CMMI) in 1999, for establishing internationally accepted definitions for mineral reserves (McKay et al, 2002).

Geoscience Australia (formerly the Australian National Geoscientific Agency) is responsible for compiling information on the nation's mineral and petroleum resources. It reports estimates of Australia's resources of all major and many minor mineral commodities on an annual basis. In doing so, it collates and analyses information gleaned from company reports to the ASX. Its classification of 'economic demonstrated resources' (EDR) includes what is currently commercial as obtained from companies' reported ore reserves and some of the reported mineral resources which Geoscience Australia considers to be economic in the longer term.

The UN-Economic Commission for Europe has developed the United Nations Framework Classification (UNFC) for reporting on Solid Fuel and Mineral Commodities resources and reserves, with particular reference to international comparisons. It was developed partly because of the need to accommodate the recent entry of the Eastern Bloc countries into the international market economy. It was released in 1997 and refined in 1998 and 1999, and is a three-dimensional system with 'feasibility', 'economic' and

Table 4.1 Some of the codes used by CRIRSCO members

Organisation, country	Name of code
CIM, Canada	CIM definition standards NI-43–101
SAIMM, South Africa	SAMREC Code
SME, USA	SME Guide for Reporting
Europe	PERC code
National Committee, Chile	Certification Code

'geological assurance' axes. It also has a numerical code for resource and reserve categories in order to overcome terminology and language problems. While the UNFC is used by governments, commercial exploration and mining companies tend to rely on one of the major country-based reporting Codes, as adopted by share markets worldwide.

As previously stated, the CMMI used the JORC Code as the basis of agreement by Australia, Canada, South Africa, UK and USA for adopting international standard definitions for mineral resources and ore (mineral) reserves in 1999. Whilst these countries have published their own revised standards, the UK has linked with the European Federation of Geologists, representing around 25 countries, and the Institute of Geologists of Ireland as PERC, to produce a reporting code for Europe. The PERC Code, which is largely based on the JORC Code but which post-dates it by five years, is likely to be the basis for developing a truly international code, which has been established and maintained by CRIRSCO. The committee currently has 13 countries as members. Some of the CRIRSCO member organisations and countries are listed in Table 4.1.

The JORC Code has been adopted by a number of international mining companies. Rio Tinto is one such company (Weatherstone, 2000). It is headquartered in the UK, with operations in around 16 countries and exploration, processing and marketing interests in many others. Its adoption of the JORC Code was based on:

- Its successful application and use in Australasia.
- Its relative simplicity.
- The excellent rapport between Rio Tinto personnel and JORC members.

Rio Tinto applauds the JORC philosophy of transparency and materiality and believes that more information is better than less, which will inevitably lead to:

- Shareholders having a better appreciation of a company's assets;
- Extending the publication of resources as well as reserves to the US in 2019, thereby giving a better picture of the company's potential;

- Uniformity of information available to shareholders regardless of their location; and
- Support from stock exchanges, financiers and analysts who are increasingly demanding higher reporting standards.

Figure 4.1 shows the key terms used in the JORC Code and their relationship. These same terms are used in almost all of the main resource and reserve reporting codes. The main difference is that the term 'ore reserve' is not used outside of Australia.

Details of the JORC Code

The key terms in the JORC Code framework are defined below in Figure 4.1.

Exploration results

Exploration results are data and information that are generated in the early stages of an exploration programme. The information should clearly indicate the method used to collect the data. These could be outcrop sampling, assays of drill hole intercepts and their sampling details, geochemical survey results and the sampling methodology (e.g. stream sediments, soil fractions and groundwater), and type of geophysical survey and results. The reporting

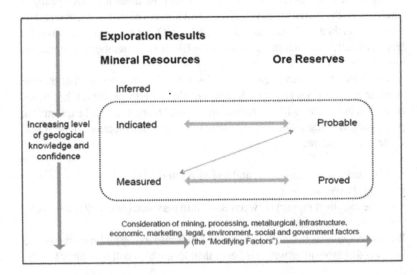

Figure 4.1 General relationship between exploration results, mineral resources and ore reserves.

of exploration results, however, must not imply that a potentially economic mineralisation has been discovered and exploration targets must be clearly identified as not being mineral resources.

Table 4.1 of the JORC Code is a checklist of the main criteria to be considered when preparing public reports for exploration companies. The table considers an initial section on sampling techniques including drilling methods and drill recovery, sub-sampling, quality of the assay data, location of data points, orientation of the data with respect to mineralisation, verification of the data and security of the samples. The reporting to criteria noted in Table 4.1 is on 'if not, why not' basis, wherein the Competent Person is required to provide a justifiable explanation to any criteria listed in the table for which no comment has been provided.

Mineral resources

Public reports on a company's mineral resources or ore reserves should include a description of the style and nature of the mineralisation. Companies must also report any significant changes in their mineral resources or ore reserves and these must be reviewed and publicly reported yearly.

A **mineral resource** is a concentration, or occurrence, of material of intrinsic economic interest in or on the Earth's crust in such a form and quantity that there are reasonable prospects of eventual economic extraction. It covers mineralisation, which has been identified and estimated through exploration and sampling and within which ore reserves may be defined. It is based on known, estimated or interpreted geological evidence and knowledge. The term 'reasonable prospects for eventual economic extraction' implies a judgement by a Competent Person in respect of technical and economic factors and approximate mining parameters. The mineral resource is further subdivided into three categories:

An **inferred mineral resource** is that part of a mineral resource for which tonnage, grade and mineral content can be estimated with a low level of confidence. It is based on information gleaned from outcrops, trenches, pits, workings and drillholes, which may be limited or of uncertain quality and reliability. Caution should be exercised if this category of mineral resource is used in economic studies.

An **indicated mineral resource** is that part of a mineral resource for which tonnages, densities, shape, physical characteristics, grade and mineral content can be estimated with a sufficient level of confidence to allow the application of 'modifying factors', leading ultimately to the declaration of an ore reserve. The information gleaned from exploration, sampling and testing is sufficient to assume geological and grade continuity between observation points where samples and data have been gathered.

A **measured mineral resource** is that part of a mineral resource for which tonnage, densities, shape, physical characteristics, grade and mineral content can be estimated with a high level of confidence. It is based on detailed and reliable exploration, sampling and testing from locations that are close enough to confirm geological and grade continuity.

Confidence in the estimates of both **indicated** and **measured** mineral resources is sufficient to allow them to be used for an evaluation of economic viability and leads to the generation of ore reserves.

The choice of category depends on the quality, quantity and distribution of the available data and is a decision taken by the Competent Person(s) authorising the mineral resource.

It should be emphasised that mineral resource estimates are not precise calculations and reporting of tonnage and grade figures should reflect their order of accuracy. For example, **inferred** mineral resources are approximate values that should be rounded off to the second significant figure.

Ore reserves

An **ore reserve** is the economically mineable part of a measured or indicated mineral resource. It includes allowances for dilution and losses that may occur due to mining. Ore reserves are the end result of appropriate assessments, which may include pre-feasibility or feasibility studies with mining, processing, metallurgical, infrastructure, economic, marketing, legal, environment, social and government factors having been taken into account (International Accounting Standards Committee, 2000).

Ore reserves are categorised as **probable** or **proved**.

A **probable ore reserve** is the economically mineable part of an **indicated**, and in some cases **measured**, mineral resource, whereas a **proved ore reserve** must be the economically mineable part of a **measured** mineral resource.

The categorisation of an ore reserve is decided mainly by the classification of the corresponding mineral resource and must be made by the Competent Person(s). Once again, it must be emphasised that ore reserve estimates are imprecise, and tonnage and grade values in public reports should be expressed appropriately.

Public reports of ore (mineral) reserves, except for coal, must clearly specify one or both categories of **proved** and **probable**, and such, categories must not be reported in combined form, unless details of the individual categories are also provided. If ore reserves are reported as containing metal or mineral content, the relevant tonnages and grades must be given.

The JORC Code provides a summary of the main criteria to be considered when preparing reports on exploration results, mineral resources and ore (mineral) reserves (Miskelly and West, 1996). For example, the following criteria should be considered when preparing reports on the estimation and reporting of ore reserves:

- Description of mineral resource estimate and a clear indication whether the mineral resources are reported additional to, or inclusive of, the ore (mineral) reserves.
- The basis of cut-off grade(s) or quality parameters.
- Mining factors inclusive of mining method(s), geotechnical factors, grade (quality) control, mining dilution and mining recovery factors and infrastructure requirements of mining method(s).
- The metallurgical process and its appropriateness, nature of metallurgical testwork, and metallurgical recovery factors.
- Cost and revenue factors, including assumptions relating to capital and operating costs, head grade, commodity price(s), exchange rates, treatment and transport charges and royalties.
- Market assessment, including demand, supply and stock situation for the commodity or commodities, commodity specifications and price and volume forecasts.
- Status of titles and approvals.
- Basis for classification of ore reserves into varying confidence categories and whether the result reflects the competent person(s)' view of the deposit.
- Results of any audits or reviews of ore reserve estimates.

When companies listed on the Australian or New Zealand securities exchanges are reporting resources or reserves, the formats of the reports are defined in guidelines published by the respective exchanges, e.g. Chapter 5 of the ASX Listing Rules, which regulates public reporting in Australia.

Australian guidelines for estimating and reporting inventory coal, coal resources and coal reserves

Before September 1999, estimating and reporting of coal resources and coal reserves in Australia was prescribed by the Australian Code for Reporting Identified Coal Resources and Reserves, as published by the Coalfields Geology Council of New South Wales and the Queensland Resources Council. The most recent version of this is the 2014 edition (Coalfields Geology Council of New South Wales, 2014).

Whereas the requirements of the JORC Code are mandatory, the procedures outlined in the Guidelines are not compulsory but are strongly recommended.

Australian Guidelines for the Estimation and Classification of Coal Reserves (2014)

As stated earlier, the JORC Code is the de facto standard for reporting of mineral resources and ore reserves and its use is mandatory for all listed entities on the Australian Securities Exchange. However, the Code acknowledges the utility of additional guidelines for estimating coal resources and reserves as published and updated from time to time by the Coal Geology Council of New South Wales and the Queensland Resources Council.

The most recent version (2014) is not part of the Code, but adherence to its processes and procedures is recommended by the Code. These Guidelines accommodate a wide range of coal deposits in rank, quality and geological environment.

It is important to note that the terms *coal resource* and *coal reserve* have the same meaning as mineral resource and ore reserve as defined in the Code.

Definition of categorisation of reasonable prospects:

- Inventory Coal is any occurrence of coal in the ground that can be estimated and reported without necessarily being constrained by economic potential, geological or other modifying factors. It is sub-divided in order of increasing geological confidence into inferred, indicated and measured (non-JORC) categories.
- Coal Resource is a realistic estimate of the coal resource under justifiable technical, economic and development conditions, which can more than likely to be classified as economically extractable.
- Coal Reserve is the economically mineable portion of a measured or indicated coal resource. It includes diluting materials as well as being adjusted for losses that may occur when coal is mined. Coal reserves are also sub-divided in order of increasing confidence into probable and proved categories.

Figure 4.2 provides an overview of the relationship between all these various categories. Australian Guidelines (2014 Edition).

Coal reserve estimates must take into account diluting materials and are adjusted for losses that may occur as coal is mined. Such assessments must also include consideration of all other relevant factors such as mining

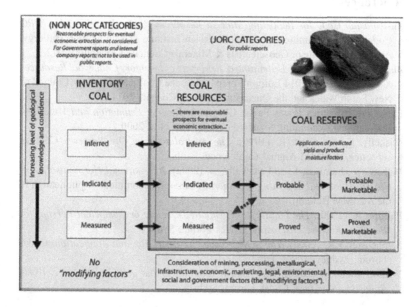

Figure 4.2 Relationships between inventory coal, resource and reserve classifications.

methods, beneficiation, economic, marketing, legal, environmental, social and other issues.

Estimation of coal reserve

Decisions regarding the category of coal reserve (probable or proved) are determined primarily by the confidence level of the assessment as determined by the Competent Person. Resource estimates should be accompanied by an assessment of the risks associated with the estimations along with the name, qualifications and experience of the estimator and his/her relationship with the tenure holder(s) and/or operators.

Marketable coal reserves are the tonnages of coal, at specified moisture content and quality available for sale after beneficiation (if available) of coal reserves, and should be reported as probable or proved marketable reserves. Finally, it must be emphasised that the Code always takes precedence over these Guidelines should there ever be any perceived conflict between the two.

References

Australasian Code for Reporting Mineral Resources & Ore Reserves (the JORC Code), Joint Committee of the Australasian Institute of Mining & Metallurgy, Australian Institute of Geoscientists & Minerals Council of Australia, 2012.

Australian Guidelines for Estimating & Reporting of Inventory Coal, Coal Resources & Coal Reserves (2014 Edition), prepared by the Coalfields Geology Council of New South Wales and the Queensland Mining Council, 2014.

Extractive Industries Issues Paper, Chapter 3, *Reserve Estimation and Valuation*, International Accounting Standards Committee, Nov. 2000.

Miskelly, N. and West, R. (1996). *The Australian Joint Ore Reserves Code in Practice*, AusIMM Annual Conference, Perth.

McKay, W., Lambert, I., and Miskelly, N. (2002). *International Harmonisation of Classification and Reporting of Mineral Resources*, submitted for publication in the AusIMM Bulletin.

Phillips, R. (2000). *The Liability of Company Directors or Competent Persons for Resource/Reserve Disclosure*, Codes Forum, Sydney.

Stephenson, P.R. (2000). *The JORC Code: Its Operation and Application*, Codes Forum, Sydney.

Weatherstone, N. (2000). *Rio Tinto's Adoption of the JORC Code as a World Reporting Standard*, Codes Forum, Sydney.

5 Mineral resources and ore reserves estimation

Introduction

Estimating mineral resources and ore reserves are among the most important tasks in the economic evaluation of prospective mines or the proposed extension of existing operations (Appleyard, 2001).

The 2012 JORC Code defines a mineral resource as *'a concentration or occurrence of solid material of economic interest in or on the Earth's crust in such form, grade (or quality), and quantity that there are reasonable prospects for eventual economic extraction. The location, quantity, grade (or quality), continuity and other geological characteristics of a Mineral Resource are known, estimated or interpreted from specific geological evidence and knowledge, including sampling. Mineral Resources are sub-divided, in order of increasing geological confidence, into Inferred, Indicated and Measured categories'*. (See Chapter 4 for details of the JORC Code etc.)

For indicated mineral resources, sample specimens and testing are too sparse to confirm geological and/or grade continuity, although their proximity is such that continuity can be assumed. A measured mineral resource, on the other hand, must have sample specimens sufficiently close to confirm geological and/or grade continuity. It follows, therefore, that tonnages, densities, shape, physical characteristics, grade and mineral content can be estimated with a reasonable level of confidence for indicated mineral resources and with a high level of confidence for measured mineral resources. The JORC Code states that only the higher categories of mineral resources can be used in evaluations of economic viability; inferred resources may not. It also states that the choice of category depends on the quality, quantity and distribution of the available data and is a decision to be taken by the Competent Person or Persons. The Competent Person must be an appropriately qualified and experienced geologist and a member of either the Australian Institute of Geoscientists or the Australasian Institute of Mining and Metallurgy.

Ore reserves are derivatives of measured and/or indicated resources and are defined as their economically mineable portions, often with the addition of diluting material. In arriving at estimates of ore reserves, therefore, appropriate allowances must be made for the inability of mining operators to extract all of the ore and their inability to do so without diluting the ore with waste rock.

These facts emphasise the subjective nature of mineral resource and ore reserve estimates, since the latter is a derivative of the former. Needless to say, the accuracy of these estimates depends in a large measure on the ability, experience and integrity of the 'competent person(s)' and those who subsequently use them in justifying executive decisions and/or attracting investors. It follows, therefore, that all basic assumptions underpinning the decisions and calculations relating to mineral resource and ore reserve estimations should be clearly stated and explained so that all users of the information can make their own judgements regarding the confidence to be placed in the estimates.

This chapter deals with the various methods that can be used to estimate the quantity and grade of mineral resources and ore reserves which are not only used to evaluate the economic viability of prospective mines but are also vital elements in the performance assessments of operational mines.

While resources and reserves are clearly critical inputs to the determination of an operating mine's financial results and balance sheets, how well understood is this?

Many inappropriate financial results have been reported by mining companies because either reserves and resources were not properly understood by people using the information or the estimates were incorrect. This issue requires attention by mining companies if errors are to be eliminated.

Before moving on to discuss the means by which mineral resources and ore reserves can be estimated, let us dwell a little on some of the assumptions that have to be made by geologists and/or engineers when converting resources into reserves.

Obviously not all of the in situ ore can be recovered regardless of how it is to be mined. Inevitably pockets of ore, particularly at the edges of the deposit, may not be recoverable, and in underground mines some ore will be sterilised in pillars to support mine workings and access/transport roadways. Consequently, the extent to which an orebody can be exploited is estimated by assuming an appropriate recovery factor.

In broad terms, the recovery factor for underground mines is usually in the range 0.7 to 0.9, except where room and pillar mining is practised, when recovery factors can vary between 0.5 to 0.9 depending on whether or not pillar recovery is carried out as a secondary phase in the extraction process.

In the case of surface mines, the extent to which the orebody can be mined is dependent on the nature and shape of the deposit (Crawford et al, 1979). For stratified planar deposits (e.g. coal seams, lateritic deposits of bauxite and sedimentary iron ore deposits) the recovery factors should be very high, say 0.9 to 0.95. On the other hand, for massive metalliferous deposits, such as nickel sulphide, porphyry, copper and disseminated gold deposits, the recovery will depend on the extent to which the mine operators can design and operate the open cut to extract the orebody. Factors such as the shape of the orebody, the ultimate depth of the open cut, the mine's stripping ratio and the angle of its final pit walls, will all influence the mine's recovery factor.

Another important issue is the extent to which dilution of the ore occurs due to the mining operator's ability to limit the extraction to just the orebody and leave the surrounding waste and low grade rock in the ground. In surface mines the overburden dilution is recorded in the mine's overall stripping ratio, defined as:

$$\text{Overall Stripping Ratio} = \frac{\text{Tonnes of waste removed}}{\text{Tonnes of ore produced}}$$

Selective mining within the open pit or opencut will, however, distinguish as clearly as possible between ore and waste so that the ore can be taken directly to the treatment plant and the waste rock to a waste dump. The waste and low grade that is unable to be separated during mining (and is treated in the treatment plant) is termed ore dilution. Ore dilution has a much bigger impact on profitability than waste dilution.

In underground metalliferous mines dilution of the ore is largely dependent on the mining method being used; following are some typical figures:

Mining method	Dilution % of ore
Cut and fill	10
Open stoping	15–20
Caving	20–30

Carras (1990) states that dilution of between 50% and 100% is not uncommon for narrow vein mining. For underground coal mining, dilution rates of between 15% and 25% are not uncommon.

Estimating mineral resources and ore reserves are, therefore, vitally important steps in the planning and development of new mines. They are the assets to be exploited and, consequently, great care should be taken to ensure they are assessed as accurately as possible. The major problem with such estimations is the paucity of data on which many of the estimates are based. Drillhole intersections provide most of the data, and the desire for plentiful good-quality information is always in conflict with the cost of drilling, which increases with the increasing depth of orebodies.

Mineral resources (a measure of mineralisation) and ore reserves (a measure of recoverable ore) are quoted as estimates of tonnages and the grade of contained valuable element or mineral.

The tonnage of ore is given by:

Volume of ore × Bulk density of ore

The volume of ore is calculated from estimates of its areal extent and thickness, which are in turn based on geological knowledge and information gleaned primarily from exploration or grade control drilling.

The grade or valuable material content of an orebody is expressed as a weight fraction and can be quoted as a percentage or as grams per tonne, etc.

The tonnage of valuable material in an orebody is therefore given by:

Average grade of ore × Tonnage of ore

It follows, therefore, that estimates of mineral resources and ore reserves are dependent on the following variables:

- Grade of ore
- Volume of ore
- Density of ore

The grade of ore is assessed by the laboratory analysis of the metallic or mineral content of the ore, whereas the volume and bulk density of the ore are based on the estimated dimensions of the orebody and measurements of the rock density respectively. All such estimates and measurements are subject to error, and great care should be taken to avoid mistakes and minimise error by keeping good quality field notes and laboratory records and double checking calculations and duplicating analyses whenever possible.

It is important at this stage to distinguish between the overall and localised ore reserves. The former relates to estimates of total resources and/or reserves to be mined over the life of the mine, whereas localised reserves refer to those portions of the orebody to be mined in the short term.

At the feasibility study stage an attempt will be made to estimate the overall resources and reserves as the asset to be exploited, in order to provide the capacity of the mine to generate an adequate cash flow to justify the initial investment of funds required for the further exploration and developmental programs.

It should also be noted that in situ resource values need to be adjusted to take account of mining recovery and dilution factors so that estimates of actual mine production can be obtained.

It is also important to emphasise the importance of grade in assessing the financial viability of a mine. Once an estimate of ore reserves has been defined to justify forward planning based on fixed production levels for 10, 15 or 20 years, any further increase in ore reserves would have no effect on production rates. It would, in point of fact, only increase the productive life of the mine.

If, however, the estimated grade of ore were to be exceeded at some point during the life of a mine, it could have an immediate positive effect on cash flow since more of the mineral commodity could be produced, assuming the mineral processing plant could cope with the increased output.

Of course, the converse is true, since a reduction in grade would have an equally immediate negative effect requiring an increase in production of ore to ensure that the forecasted cash flow is maintained. This assumes that ore production can be increased and that the mine transport system and processing plant are able to cope with such an increase in production.

Definition of terms

Grade: The grade of an orebody is its valuable mineral or element content expressed as a weight fraction. It can be quoted as a percentage or as grams per tonne, parts per million, etc.

             It can refer to the metal or mineral content of the ore, with the latter being the most important if it contains the valuable element.

Range: The valuable material content in a mineral deposit depends on the price the metal/mineral can be sold for and can vary from as little as few parts per million for gold and diamonds, to several percent for copper and almost 100% for coal.

             The range of valuable material within some types of orebodies can vary significantly. The gold mining spawned the terms 'mother lode' and 'bonanza' because gold ore grades can range from one quarter to one tenth normal grade, to rich zones 1,000 to 10,000 times the normal grade.

             For copper, deposits average 1% to 5% copper, but the range of samples can extend from low values of, say, one tenth normal grade to high values of 10 to 20 times normal.

Iron ore deposit average grades are about 60% iron, with a much tighter range, extending from one tenth below to very little above normal grade.

The practicalities of mining require mine operators to define the lowest limit of profitability or cut-off grade (COG) for every orebody. The COG is also used in reporting mineral resources.

Cut-off grade:
The JORC Code requires companies to quote ore reserves as the part of mineral resources that are economically recoverable, often with the addition of diluting material (not in the resources). This involves defining the lowest grade of ore that can be economically exploited, which is known as the cut-off grade.

In the limit, when mining ore at the cut-off, earnings must equate to total costs. Thus for a gold ore we have:

$$COG = \frac{Costs\,(\$\,per\,tonne)}{Earnings\,(\$\,per\,gram)} \times \frac{1}{RF}$$

Where RF = Processing plant recovery factor

A mining recovery factor is also often added.

As an example assuming that the gold price is US$1,300 per oz, the operating and processing costs are US$50 per tonne and the mill recovery factor is 95%, the COG is given as (Note: 1 troy ounce = 31.1 grams):

$$COG = \frac{50}{1300} \times \frac{31.1}{0.95} = 1.3\,grams\,per\,tonne$$

Metal/ mineral distribution:
Rogers and Adams (1963) demonstrated that for ores with an element or mineral content in the region of 20% to 80%, the material will tend to be normally distributed in homogeneous rock. They also found that as the concentration approaches 100%, the distribution becomes negatively skewed, whereas as it approaches zero the distribution is often lognormal.

Specific gravity:
This dimensionless number is obtained by dividing the density of a substance by the density of water, which is 1,000 kg/m^3 or 1 tonne/m^3.

Density:
The density of a substance is its mass per unit volume (Lipton, 2001).

Dry bulk density:
The effective dry density of a parcel of rock measured in mass per unit volume including effects of porosity, cracks, fractures and caverns.

In estimating mineral resources and reserves, it is the dry bulk density (i.e. mass per unit volume excluding any natural water) that should be used.

Nevertheless, it is the (in situ) bulk density (i.e. dry bulk density plus natural water content) that should be used when estimating the tonnage of material to be transported and processed.

The density of ore is never uniform but varies significantly throughout the orebody in the same way the grade does. The density must be estimated from hard data wherever possible. A significant number of density measurements should be determined for each rock type.

Getting the density estimate incorrect by 20% has the same disastrous effect on mineral economic assessment as getting the grade wrong by 20%. Recall the equation for contained metal in a deposit:

Metal = Grade × Volume × Bulk Density

The simplest method of determining the density of a rock sample is the so-called caliper method, which can be used if good-quality drill core is available. Knowing the diameter of the specimen, a, its length, I, and weight after drying, M_d, the volume, V, and dry bulk density, P_d, can be calculated as follows:

$$V = \left(\frac{\pi d^2}{4} \times l \right) m^3$$

$$Pd = \frac{Md}{X} kg/m^3$$

Weights assigned to observations: Standard error, or mean square error, σ, is the preferred measure of the precision of an observation or observations. For a single observation:

$$\sigma_{(s.o)} = \sqrt{\frac{\sum (x - \bar{x})^2}{n-1}}$$

And for the arithmetic mean:

$$\sigma = \sqrt{\frac{\sum (x - \bar{x})^2}{n(n-1)}}$$

If all measurements of a variable were deserving of the same emphasis, the most probable value of the variable would be its arithmetic mean. If, on the other hand, the measurements were deserving of varying emphases, the most probable value would be given by the weighted arithmetic mean.

Assume that a survey line was measured six times on one day and three times the following day, under identical conditions and using the same techniques.

It can be assumed, therefore, that the standard error of a single observation is the same on both occasions. It also follows that the standard error of the means is given by:

$$\sigma_{Day1} = \frac{\sigma(s.o)}{\sqrt{6}}$$

$$\sigma_{Day2} = \frac{\sigma(s.o)}{\sqrt{3}}$$

In this case the mean of the measurements on Day 1 should be afforded more emphasis than the mean of the measurements on Day 2 when calculating the most probable length of the survey line. It also follows that since the standard error of all individual measurements can be assumed to be the same, the means for each day should be weighted according to the number of measurements taken. Thus, the most probable value of the length of the survey line, L, is given by:

$$L = \frac{^6 Z_1 + ^3 Z_2}{6+3} = \frac{W_1 Z_1 + W_2 Z_2}{W_1 + W_2} = \frac{\sum(WZ)}{\sum W}$$

Where Z_1 and Z_2 are the means of the first and second set of measurements and W_1 and W_2 their corresponding weights.

It also follows from the expressions for the standard error of the means of both sets of measurements that:

$$\frac{W_1}{W_2} = \frac{6}{3} = \frac{(\sigma_2)^2}{(\sigma_1)^2}$$

Thus the weights are inversely proportional to the square of the errors.

Methods of estimating mineral resources and ore reserves

Resources and reserves are quoted as estimates of tonnages and grade and the available methods fall into one of three main categories:

- Classical methods
- Distance weighted methods
- Geostatistical method

The first two methods are no longer used in the mining industry for public reporting of resources and reserves, but are useful for 'back of the envelope' estimates when a quick result is needed. They also are important for teaching the concepts of estimation to students.

Classical methods

These methods are based on simple and easily understood principles.
The methods that fall into this category can be classified as follows:

- Block method
- Polygonal method
- Cross-sectional method

The blocks may be rectangular or triangular, with drillholes defining the orebody at each corner.

Rectangular blocks

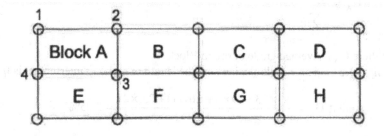

Let us assume the shallow drillholes defining the orebody in Block A provide the following information:

Drillhole	Ore intersection (metres)	Grade % Copper	Distance between holes (metres)
1	23	2.1	
			100
2	29	3.0	
			70
3	33	1.6	
			100
4	27	2.4	

Also assume that the drillholes have the same collar elevation so that the upper surface of the orebody is more or less level.

Thus, for Block A:

Area of block = 7,000 m^2
Average thickness of ore = 112/4 = 28m
Volume of ore = 196,000m^3

Given that the specific gravity of the ore determined from drillhole specimens is 2.8, we have that:

Tonnage of ore = 548,000 tonnes

In calculating the average grade of ore for the block, it is appropriate to place more emphasis on those grades representing the thicker intersections, since the number of specimens used for the analysis of ore grade is directly proportional to the length of the intersection. Thus:

$$G_A = \frac{\sum\limits_{i=1}^{4} Gi_i \, l_i}{\sum\limits_{i=1}^{A} l_i}$$

Where G_A = Average grade of ore in Block A
Gi_i = Ore grade for drillhole, i, where the length of orebody intersection is I_i

so that $G_A = \dfrac{2.1\times23+29\times3+1.6\times33+2.4\times27}{12} = 2.25\%$

Consequently, the metal content in Block A is given by:

548,000 × 2.25/100
= 12,330 tonnes

The total tonnage of ore in all blocks is given by the sum of the tonnages calculated for each block.

In this case, since the areas of all blocks are the same and assuming the specific gravity of the ore is more or less constant, the mean grade for all ten blocks would be the arithmetic mean of the grades calculated for each block.

Triangular blocks

If the drillholes are not uniformly spaced, it may be appropriate to split the orebody into a series of triangular blocks.

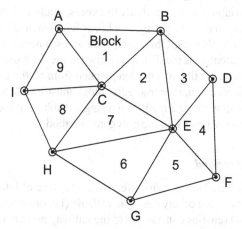

Assuming, once again, a level orebody and a constant specific gravity, for Block 1:

$$\text{Average thickness of ore} = \frac{\sum\limits_{i=1}^{n} t_i}{n} = \frac{t_A + t_B + t_c}{3} = \bar{t}$$

Where t_A, t_B, t_c = thickness of ore at drill-holes A, B and C respectively
$\bar{t}$ = average thickness of ore for Block 1.

$$\text{Average grade of ore} = \frac{\sum\limits_{i=1}^{n} (G_t)}{\sum\limits_{i=1}^{n} t} = \frac{G_A t_A + G_B t_B + G_c t_c}{t_A + t_B + t_c} = G_1$$

Area of Block = ½ triangle base × height
        = A_1
Volume of ore = $A_1 \times \bar{t} = V1$
Tonnage of ore = $V_1 \times SG$

For all mine blocks, the total tonnage of ore is the sum of the tonnages calculated for each of the triangular blocks.

In this case, the mean grade of ore for the orebody contained within this network of drillholes would be given by the area weighted mean, i.e.

$$\bar{G} = \frac{\sum\limits_{i=1}^{n} G_i A_i}{\sum\limits_{i=1}^{n} A_i}$$

The rectangular and triangular block methods are based on the assumption that the variables such as orebody thickness, grade and specific gravity, change gradually and continuously as a linear function of distance along the lines joining the drillholes. Obviously, such an assumption becomes more and more questionable as the distances between the drillholes increase.

Nevertheless, using the information gleaned from all the drillholes at the corners of each rectangular or triangular block must lead to a better assessment of the ore than that gleaned from a single drillhole at the centre of the polygonal blocks, when using the polygonal method.

The polygonal method

This method is based on the principle of 'equal sphere of influence', which assumes that the value of any regional variable (i.e. ore thickness, grade or specific gravity) remains constant up to the halfway point between adjacent drillholes.

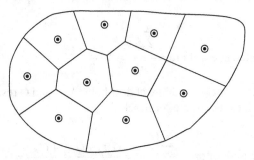

The polygons are formed by the perpendicular bisectors of the lines joining adjacent drillholes. Consequently, each drillhole is located within its own polygon and it is assumed that the assay values (i.e. thickness, grade and specific gravity) relating to each drillhole pertain to the entire ore contained within the polygonal block.

The mechanics of the method can be defined as follows:

1   Lay out the polygons by drawing perpendicular bisectors of the lines joining the drillholes. The outer polygons are made equidimensional around the drillhole.
2   Measure the area of each polygon and multiply it with the thickness of the ore at the drillhole to obtain the volume of ore within the polygon. Thereafter the tonnage of ore is calculated by multiplying the volume of the polygonal block with the ore's specific gravity.
3   The total tonnage of ore in the orebody is given by the sum of all the polygonal tonnages.

4   The mean grade of the overall deposit may not be particularly meaningful, because it often relates to a fairly extensive deposit with relatively few drillhole assays. Nevertheless, when it is calculated it should be done using an appropriate weighting method.

Obviously the assumption that the grade of the ore in each block is the grade determined from the central borehole specimen assays will become less accurate as the area of the polygon increases. Consequently, the weighting of polygonal block grades should reflect this fact so that higher weights are accorded those blocks with smaller areas, i.e.:

Weights $\times 1 / (\text{Area})^n$

Whereas the value of the exponent, n, is unknown, it would be appropriate in the lack of further information to assume it is unity since it is claimed that for many orebodies drillhole values should be weighted according to the inverse of distance squared, i.e.:

Weights $\times 1 / d^2 \times 1 / A$

It therefore follows that the mean grade of the orebody, when using the polygonal method of ore reserve estimation, is given by:

$$\bar{G} = \frac{\sum_{i=1}^{n} \left(G_i x \frac{1}{A_i} \right)}{\sum_{i=1}^{n} \left(\frac{1}{A_i} \right)}$$

If the drillholes are laid out in a rectangular pattern the polygons become rectangles.

⊙	⊙	⊙	⊙
⊙	⊙	⊙	⊙
⊙	⊙	⊙	⊙

There is a common problem with all methods of resource or reserve estimation and that is how best to deal with the outer limits of the orebody. Geologists will make their best estimate of the outer limits, both in regard to their areal and crosssectional extents, and all ore reserve estimates must relate to these limits.

In the polygonal method, as stated earlier, the outer limit polygons are made as equidimensional as possible around the drillholes. With the other methods, the estimators have to make reasonable and appropriate assumptions in order to ensure the outer limits are included in the overall appraisal. This entails calculating the ore in unusually shaped blocks based on the analysis of nearby drillhole assay data.

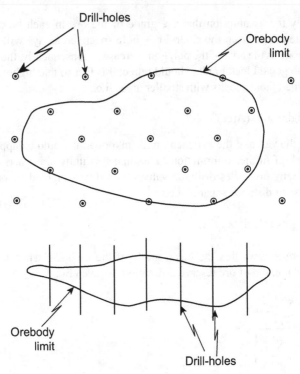

Cross-sectional methods

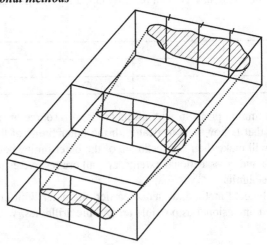

In this instance the drilling is organised to provide information regarding specific cross-sections at set distances apart, as demonstrated in the diagram. The subsequent calculations can be approached in one of two ways, depending on whether we adopt the principle of gradual change or the principle of equal sphere of influence.

Using the principle of equal sphere of influence we split the orebody into blocks, e.g.:

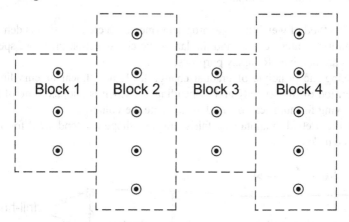

Each block extends halfway to the next series of drillholes.

For Block 1:

$$\text{Mean grade} = \frac{\sum_{i=1}^{n}(G_i \cdot t_i)}{\sum_{i=1}^{n} t_i} = G_1$$

And likewise if the SG varies.

$$\text{Mean grade} = \frac{\sum_{i=1}^{n}(SG_i \cdot t_i)}{\sum_{i=1}^{n} t_i} = SG_1$$

$$\text{Volume of block} = \text{Orebody cross-sectional area x distance between cross-sections}$$
$$= A_1 \times I$$
$$\text{Tonnage of ore in block} = A_1 \times I \times SG$$

The mean ore grade for the deposit as a whole should be the volume or tonnage weighted mean depending on whether or not the orebody has a relatively constant specific gravity.

The tonnage weighted mean ore grade is given by:

$$\text{Mean grade for the deposit} = \frac{\sum_{i=1}^{n}(G_i x Tons_i)}{\sum_{i=1}^{n}(Tons_i)}$$

Where n = number of blocks

This method of weighting is appropriate since each cross-section is defined by multiple intersections, and the larger the cross-section the more specimens are obtained for assay purposes.

Using the principle of gradual change leads to a linear or curvilinear definition of the orebody and, depending on which is adopted, one of the following formulae can be used to calculate the volumes.

If the orebody maintains a fairly uniform shape the 'end-area' formula can be used.

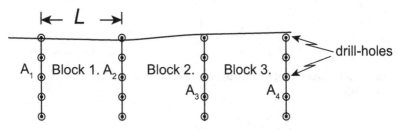

For Block 1: Volume = $(A_1 + A_2)/2 \times L$

If, on the other hand, the orebody changes in shape, such that it is best defined by a curvilinear surface, the primordial formula will give the best result, i.e:

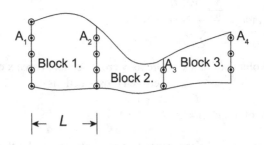

For Block 1:   Volume = $(A_1 + 4A_m + A_2)\,L/6$
Where A_m = mean area between sections
L = distance between sections

Other formulae of use include:

The 'wedge formula' for those sections of an orebody that taper to a thin line.

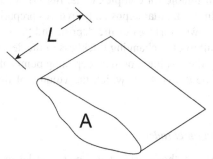

Volume = A/2 × L

And the 'cone' formula for those portions of an orebody that taper to a point.

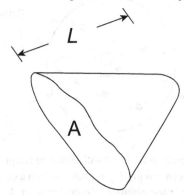

Volume = A/3 × L

In this case the ore grade for each block would be the thickness weighted mean of all the specimen ore grades, i.e. for Block 1:

$$G_1 = \frac{\sum_{i=1}^{n}(G_i t_i)}{\sum_{i=1}^{n} t_i}$$

Where n = number of orebody sections
G_i = mean ore grade for Block 1
t_i = thickness of intersection

Once again, the ore grade for the entire deposit would best be determined by the volume or tonnage weighted mean, depending on whether or not the specific gravity is relatively constant throughout the orebody.

Contouring

This method is unsuitable for complex metalliferous deposits but is nevertheless quite useful for tabular deposits. It involves preparing contour maps of the upper and lower surfaces of the deposit and the superimposition of one on the other allows the changing thickness of the deposit to be defined.

From these thickness values contours depicting points of equal thickness, or isopacks, can be drawn from which the volume of the deposit can be calculated.

Distance weighting methods

Distance weighting methods were developed in order to provide a means of estimating the grade of un-assayed blocks or points from surrounding data, e.g.:

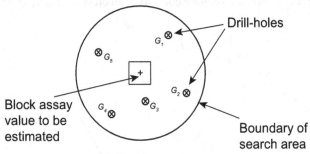

The underlying principle associated with distance weighting methods is that a correlation exists between grades at points within a certain distance of one another, and the nearer the points are to one another the stronger the correlation. This is a logical assertion which has proved to be the case for many orebodies, and the major questions it raises are:

1   What is the nature of the correlation, and
2   Over what distance does the correlation exist?

Since it is assumed the correlation between grade values weakens with distance, it follows that some form inverse distance weighting could apply, i.e.:

Weighting for assay values $1/d^n$

Where d = distance between the point/block whose grade is sought and
drillhole having a known grade
n = exponent of the distance

One method of deducing the value of n is to estimate the grade of each assay value in a group of known values based solely on the other values, using different values of n, and to record the difference between the estimate and known value (i.e. the error) in each case. The lowest accumulated error for the group of assay values will indicate the most appropriate value of n for that orebody.

The estimated grade is, in each case, obtained using the following equation:

$$G = \frac{\sum_{i=1}^{n}\left(G_i \cdot \frac{1}{d_i^n}\right)}{\sum_{i=1}^{n} 1/d_i^n}$$

The most common values of n are either 1, 2 or 3 which relate to the inverse distance (ID), inverse distance squared (ID^2) and inverse distance cubed (ID^3) methods respectively.

The formulae associated with these methods are:

Inverse Distance:

$$G = \frac{1/d_1 \, G_1 + 1/d_2 \, G_2 + \ldots\ldots + 1/d_n \, G_n}{1/d_1 + 1/d_2 + \ldots\ldots 1/d_n}$$

Inverse Distance Squared:

$$G = \frac{1/d_1^2 \, G_1 + 1/d_2^2 \, G_2 + \ldots\ldots + 1/d_n^2 \, G_n}{1/d_1^2 + 1/d_2^2 + \ldots\ldots 1/d_n^2}$$

Inverse Distance Cubed:

$$G = \frac{1/d_1^3 \, G_1 + 1/d_2^3 \, G_2 + \ldots\ldots + 1/d_n^3 \, G_n}{1/d_1^3 + 1/d_2^3 + \ldots\ldots 1/d_n^3}$$

In using any of these methods, it is necessary to identify the search area and have a balanced spread of assay values, which may mean excluding some that are too closely in line with others, e.g.:

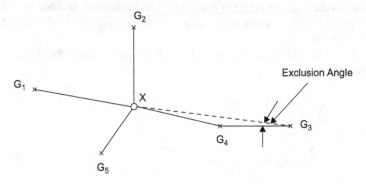

Assay values G_2 and G_4 are two such values and G_3 should be eliminated in order to avoid placing too much weight along that trend line in estimating the grade at X. Such eliminations can be made by adopting an 'exclusion angle' (e.g. 20°) so that the outer of two values having an exclusion angle less or equal to the quoted value is eliminated.

Example: Given the following gold assay data with values quoted in parts per million (ppm) calculate the grade at the centre of the unassayed block using all three inverse distance methods.

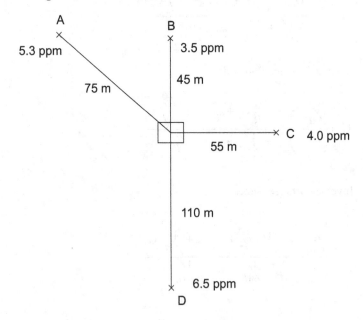

Solution:

Weights

Points	$1/d$	$1/d^2$	$1/d^3$
A	0.013	0.00018	2.3×10^{-6}
B	0.022	0.00049	10.8×10^{-6}
C	0.018	0.00033	5.95×10^{-6}
D	0.009	0.00008	0.7×10^{-6}
Σ weights	**0.062**	**0.00108**	**19.75×10^{-6}**

Weighting factors

Points	$\dfrac{1/d_i}{\Sigma(1/d_i)}$	$\dfrac{1/d_i^2}{\Sigma(1/d_i^2)}$	$\dfrac{1/d_i^3}{\Sigma(1/d_i^3)}$
A	0.21	0.17	0.12
B	0.35	0.45	0.55
C	0.29	0.31	0.30
D	0.15	0.07	0.03
Total	**1.00**	**1.00**	**1.00**

Method	Calculated assay value
ID	4.48 ppm
ID^2	4.18 ppm
ID^3	3.97 ppm

To satisfactorily apply any of the distance weighting methods it is necessary to know or to determine:

1 Whether the orebody has an isotropic (similar in all directions) or anisotropic (the opposite of isotropic) mineral structure;
2 The shape (i.e. circular, elliptic, square or rectangular) and extent of the search area; and
3 The value of the exponent, n.

Unless the range of influence for the orebody has been defined geostatistically by calculating a variogram or semivariogram (see below), the person responsible for the estimations has to make a judgement based on experience and information available regarding the shape and extent of the search area.

Experience has, for example, demonstrated that porphyry copper style orebodies can usually be adequately defined using the ID^2 method.

It is also worth bearing in mind that the choice of the exponent is essentially arbitrary and cannot be calculated from any feature of the data.

It is also the case that estimators sometimes use an inverse distance weighting method as a check on the results of a geostatistical grade interpolation method such as the kriging process.

Geostatistics

Geostatistics is based on the theory of regionalised variables developed primarily by Matheron (1970). It is based on the assumption that the statistical distribution of the difference in grade between pairs of specimens is similar throughout the deposit or zone to be estimated, being dependent solely on the distance between and the orientation of the pairs of specimens.

Ore grade is a regionalised variable and if it is possible to examine it in sufficient detail within an orebody, geostatistical methods can be used to define the resource. Such methods can lead to optimal solutions but only if a substantial quantity of good quality data is available.

Unfortunately, only limited data is usually available at the initial exploration stage and, consequently, meaningful semivariograms can rarely be obtained and fitting models to such sparse data is unwise. Once infill drilling has been undertaken geostatistical methods can be applied effectively.

The geostatistical resource estimation process involves the following steps:

- Load all drillhole geology and assay data into a mining software (such as Micromine™ or Surpac™) and interpret the geology of the deposit in 3-dimensions (3D). A set of 'wireframes' is created and used to domain all samples into zones of similar geology/grade characteristics.
- Validate the laboratory data using QAQC results and statistical analysis. Poor-quality data is removed.
- Use drillhole and/or other in situ data to develop a semivariogram or semivariograms for each and every domain.
- Fit a model to the semivariogram(s).
- Use the semivariogram(s) to determine the range of influence so that a search area can be defined for each domain.
- Build a 'block model' of the orebody by developing a computer generated 3-dimensional grid of blocks to which attributes are assigned such as, X-Y-Z coordinates, X-Y-Z size, rock type, weathering type, ore domain.

- Use the selected grade interpolation method (e.g. kriging) to estimate the grade of each and every block of the block model within the ore-body. Typically, drill samples from each domain are used to interpolate grades into blocks of the same domain. Input values and orientations obtained from the semivariograms are inserted into the kriging equations, with each block having a separate set of values. Separate estimates are made for each metal, and of critical contaminant constituents (e.g. SiO_2, Al_2O_3, P, LOI for iron ore).
- Values for bulk density are assigned based on the density data. This is usually done by averaging data for each domain but can be estimated from the 3D locations if sufficient density samples have been measured.
- Validate the resultant block model by using statistical comparison and visual comparison of drill samples grades to model cell grades, using 3D mining software.
- Calculate the volume, tonnes and metal in each block in the block model. Total all the blocks with a grade above the cut-off grade to determine the total tonnes and metal in the deposit.
- Classify the blocks into the three resource categories of inferred, indicated and measured, based upon geological and grade confidence. Total all the blocks with a grade above the cut-off grade to determine the total tonnes and metal in each category.

The variogram/semivariogram

One of the fundamental tools of geostatistics is the variogram or semivariogram. It is a graph demonstrating the extent to which the ore grade within an orebody is dependent on distance.

Ore grade variance is related to the difference in grade between specimen values according to the following equation for the variogram:

$$2\gamma = \sum_{i=1}^{n} (G_i - G_{i+1}) \frac{2}{n}$$

Where γ = variance of sample values
   N = number of specimen values
   G_i = grade at point, i
   G_{i+1} = grade at point, i + 1
   L = distance between specimens.

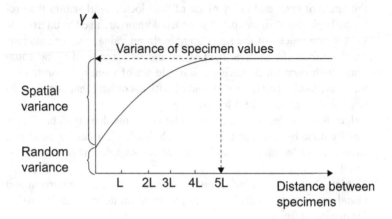

It is logical to assume that the grade at points closer together should be more related than the grades at points further apart. The variance of the distribution of differences should also be low if the values are highly correlated and vice versa. Consequently, this variance can be used as an indicator of the correlation between samples and when plotted in graphical form it is known as the variogram if $2\gamma(I)$ is used and the semivariogram if $\gamma(I)$ is plotted against the distances between specimens. Typically in the 21st century, estimators always generate a semivariogram, but this is referred to informally as a variogram.

Let us assume the following gold grades in grams per tonne (g/t) were obtained from the analysis of drill-core specimens collected from a line of drillholes, 10 m apart:

2   3   5   6   7   8   7   7   5   4   2   4

The arithmetic mean of these values is 5 g/t.

The variance of these specimen values is given by:

$$\frac{\sum(\text{Mean value} - \text{Single value})}{\text{Number of samples}} = 3.8$$

At one drillhole interval (i.e. 10 m) the variance is given by the average squared difference between specimen values, i.e.

$$\frac{(2-3)^2+(3-5)^2(5-6)^2+(7-8)^2+(8-7)^2+(7-7)^2+(7-5)^2+(5-4)^2+(4-2)^2+(2-4)^2}{11} = 2.4$$

For two drillhole intervals (i.e. 20 m) the variance is:

$$\frac{(2-5)^2+(3-6)^2+(5-7)^2+(6-8)^2+(7-7)^2+(8-7)^2+(7-5)^2+(7-4)^2+(5-2)^2}{10} = 4.6$$

The values for 3 and 4 drillhole intervals are 9.6 and 17 respectively, demonstrating their tendency to increase in value as the drillhole intervals increase.

The variogram or semivariogram based on plotting $2y(I)$ or $y(I)$ respectively can provide valuable information about the orebody including:

- The 'range of influence' which is the distance within which there is a correlation between grades, and
- The extent to which the orebody exhibits a 'nugget' effect.

Once the range of influence is determined for a specific orebody, it should be used to determine the preferred density of drilling, since drillhole spacing should ideally be within 0.7 to 0.85 of this range.

Needless to say, in an isotropic orebody (something which is very rare), the range of influence would be the same in all directions, but would be directionally dependent in an anisotropic deposit (by far the most common scenario). For instance, in a bedded type iron ore deposit, the range along strike could be as long as 200 m but in the across bedding direction could be as short as 10 m.

If, for example, the range of influence is 160 m in the N to S direction and 100 m from E to W, drillhole spacings should ideally be about 130 m N to S and 80 m E to W.

The range of influence can vary greatly from one orebody to another. It can be as little as 10m or so in a gold quartz vein and in excess of 100 m in a base metal deposit.

The nugget effect or random variance can be described as the difference in assay values between the two halves of a split drill-core specimen. This local, random variance is relatively small in most cases but can be large in gold orebodies.

If the nugget effect is large, kriging may not apparently provide a better estimation than other methods, but when the nugget effect is small geostatistics will provide the best result. That apart, kriging will also provide a measure of the accuracy of estimations, something which no other method provides.

If that is the case, the practitioner should use geostatistics whenever it is appropriate to do so. Most mineral resource estimates in Australia are undertaken using 3D mining software and geostatistical analysis. Ordinary kriging is the most commonly used grade interpolation method. ID^2 and ID^3 are sometimes used for inferred resources and only as a check estimator for indicated and measured resources. Other interpolation methods used include simple kriging, co-kriging, disjunctive kriging and indicator kriging.

It must always be borne in mind, therefore, that a highly erratic set of grade values indicating an apparent lack of correlated mineral structure (i.e. a pure nugget effect) may well be due to the fact that the specimens were obtained from points too widely spaced within the orebody.

That apart, the precise nature of the correlation can vary from one orebody or zone to the next. This needs to be modelled using one of a small number of mathematical functions so that weights can be estimated in every direction and at every distance. Some of the models used are the popular 'Matheron' or spherical model, and the so-called exponential or linear models. These models are purely mathematical representations of the 'experimental' semivariograms, based upon the data, and imply no genetic relationship to the mineralisation.

Theory of semivariograms

There are a number of theoretical models that can be used to represent and model the experimental semivariograms, a practice known as structural analysis. They can conveniently be placed into two groups:

Group 1: Models without sills

Generalised linear $\gamma(\ell) = \mathrm{m}.\ell.^x \; x < 2$

De Wijsian $\gamma(\ell) = 3\log_e \ell$

Group 2: Models with sills

Spherical $\gamma(\ell)$ $= C\left[\dfrac{3l}{2a} - \dfrac{\ell^3}{2a^3}\right] b < a$

$= C \qquad\qquad I > a$

Exponential $\gamma(\ell)$ $= C\left(1 - \exp \dfrac{l}{a}\right)$

Nugget effect $\gamma(\ell)$ $\begin{array}{ll} O & \ell = O \\ C_o & \ell > o \end{array}$

The models in Group 1 are those in which the values of $\gamma(\ell)$ increase continuously with distance. The simplest and most popular is the linear case where $\gamma(\ell) = m\ell$ with m being the gradient of the line. Other so-called linear cases depicated by $\gamma(\ell) = m\ell^x$ involve the distance, ℓ, being raised to some power, x, that lies between O and 2. The only other model in this group is the De Wijsian model which has $\gamma(\ell)$ changing linearly with increasing values of $\log_e \ell$.

To recap, the semivariogram, γ, is one half of the variance of the mean squared difference in grades for specimens separated by multiples of a set distance apart, b. Consequently, if we have 20 specimens from the mineral deposit at equal distances apart, we have that:

$$\gamma(1) = \frac{1}{2} \times \frac{1}{19}\left[(G_1 - G_2)^2 + (G_2 - G_3)^2 + \ldots\ldots + (G_{19} - G_{20})^2\right]$$

$$\gamma(2) = \frac{1}{2} \times \frac{1}{19}\left[(G_1 - G_3)^2 + (G_2 - G_4)^2 + \ldots\ldots + (G_{18} - G_{20})^2\right]$$

etc.,

so that in general we have that:

$$\gamma(b) = \frac{1}{2(\pi - h)} \sum_{i=1}^{n-h}(G_i - G_{i+h})^2$$

where G_i = grade at each point

  n = number of specimens

  h = 1, 2, 3, ..., 20

Each of these values becomes a point on the graph (i.e. semivariogram) which defines the relationship between y and multiples of h for a given direction.

As stated earlier, knowing the range of influence, which may be defined by the semivariogram, can be extremely useful to geologists and mining engineers.

It can help decide the most appropriate distances apart for the exploration drillholes. Specimens of ore may also be collected along the levels (i.e. tunnels which are more or less level) delineating the upper and lower limits of future stopes. If, for example, the distance between these levels is 40 m and the range of influence is 10 m, then no matter how many specimens are collected along the upper and lower levels they will tell us nothing about the central 20 m. Consequently, only inferences can be made regarding the nature of the deposit in this region. Such situations are not uncommon in gold quartz veins but are less likely in base metal deposits, where the range of influence is usually much larger and may be in excess of 100 m (Royle, 1980).

It is also important to remember that geostatistics is based on probabilistic principles, such that it assumes that the distribution of the differences in grade between point specimens the same distance apart and orientation is the same all over the deposit or within each estimation domain. Such distributions are usually defined by their means and variance. Whereas the mean demonstrates the trend in grade values over the deposit, the distribution's variance provides a measure of the inter-dependence of the grades between two points separated by a known vector distance, *h*.

In practice, as Clark (1982) states, those deposits which are essentially two dimensional in character, such as vein or sedimentary deposits with a reasonably constant thickness at fairly shallow depths, can be explored by fairly close drilling (e.g. 50 m to 75 m grid with closer drilling in a smaller area). This allows semivariograms to be drawn in many directions so that the deposit's isotropy or (most usually) lack thereof can be determined.

Nevertheless, many exploration programmes, particularly those relating to the deeper deposits, involve drilling at regular intervals in those directions of primary interest but usually at considerable distances apart. In such instances, good semivariograms can be obtained in down-the-hole directions, but very little information will be gleaned about ore grade correlations in the horizontal or lateral directions. However, every effort must be made to gather as much good-quality data as possible from the mineral/ore deposit at minimum cost and perhaps the only method currently available for the deeper deposits is that of deflecting or wedging the holes in order to get additional intersections from selected holes.

If insufficient data is available to satisfactorily define the nature of the ore grade variations within the deposit, the exploration team has no alternative but to fall back on its knowledge of the geology and structure of the deposit. Once a decision has been made to develop a mine, however, its development headings and declines will provide exposures of ore from which closely spaced specimens can be obtained, thereby allowing good-quality semivariograms to be constructed in various directions for grade control estimation.

Geostatistics should, therefore, never be seen as an alternative to geological investigations but rather as an extremely valuable complementary tool. If, as stated earlier, the exploration team knows that the range of influence is greater in one direction than another, the drillholes can be spaced accordingly. It is also sensible to concentrate the exploration tasks more intensively in those areas of the orebody to be mined first.

The type of drilling to be used is also dependent on a number of factors, including the depth of the deposit, the nature of the information required and the costs involved. As stated by Call (1979), when exploring for uranium, the primary need is for down-hole geophysical logging so that much of the drilling can be done using low-cost non-coring methods. If, on the other hand, good-quality geological information is required nothing can improve on diamond coring unless the rock is fractured and core recovery is poor, when it is better to use one of the other lower-cost methods such as reverse circulation drilling.

Geostatistical techniques for estimating ore grades

As stated above, the semivariogram, γ, is a graph demonstrating how grade differences in a mineral deposit varies with multiples of a given distance between specimens in a particular direction. If it is the same in all directions, then the orebody is said to be isotropic, a very rare occurrence in mining. The variance of these grade differences can be defined as follows:

$$\gamma(h) = \frac{1}{2N} \sum_{i=1}^{n} (G_i - G_{i+h})^2$$

where N = number of grade differences in the sample

Semivariogrms indicate how the rate of growth of $\gamma(h)$ occurs with increasing values of h. This can vary from a gradual rate of growth from zero to a clearly defined 'sill' value, as often occurs in a sedimentary deposit, to highly irregular changes over short distances as commonly occurs in some gold deposits. If evidence of continuity or correlation is totally non-existent, the semivariogram will be a straight line, demonstrating a pure 'nugget' effect, when specimen grade values are independent of one another. Indeed, if drillhole spacing is beyond the deposit's range of influence, the semivariogram will also demonstrate a pure nugget effect, indicating a deposit with random fluctuations of grade. In this case there is, of course, an underlying cohesive structure, which cannot be revealed due to the wide spacing of drillhole specimens, as shown in the following diagram.

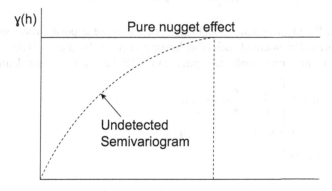

Distance between specimens

Not all semivariograms demonstrate a range of influence and a sill. Some deposits demonstrate γ(h) increasing continuously with increasing values of h. David (1977) states that in most hydrothermal deposits their semivariograms are a straight line when γ(h) is plotted against ln(h), such that:

$$\gamma(h) = a\,h^{\lambda}$$

and $\gamma(h) = A\,\ell n\,h + B$

where A and B are constants
and $\ell nh = \log_e h$.

This is, of course, the previously mentioned De Wijsian model.

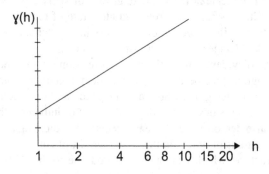

Idealised De Wijsian Semivariogram

By far the most popular model that occurs in those deposits where grades become independent of each other beyond a certain distance, a, is the Matheron or (more commonly) the spherical model. Let us once again define it:

$$\gamma(h) = C\left[\frac{3}{2}\frac{h}{a} - \frac{1}{2}\frac{h^3}{a^3}\right] + C_o\,.\,h \leq a$$

$$\gamma(h) = C + C_o \qquad\qquad h > a$$

$$\gamma(o) = o$$

where   a = range of influence
         C_o = nugget effect (if present)
      $C + C_o$ = sill value

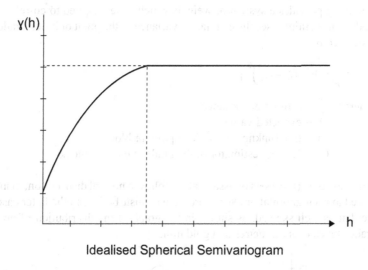

Idealised Spherical Semivariogram

The spherical model has been successfully applied to many sedimentary and other deposits including iron ore, porphyry copper, lead and zinc, and lateritic bauxite and nickel deposits (David, 1977).

Kriging

Kriging is a geostatistical procedure by which the 'best linear unbiased estimator' (BLUE) of the unknown grade of a point or a block of ore can be deduced. The word was brought into use by Matheron in France in about 1960, and was named after D. G. Krige who was among the first to make use of the technique in the evaluation of mineral resources. The method provides a procedure for calculating the most probable values of the weights to be allocated to the grade values located within the sphere of influence of the point or block whose grade is to be estimated.

As Brooker (1980) states, local estimates of ore grades at points or blocks in a mineral deposit are often carried out without due regard to the variability of the grades in the deposit. The polygonal method, for example, assumes the grade of the central specimen to be truly representative of the grades throughout the polygonal block. The inverse-distance (ID) method, on the other hand, acknowledges the fact that correlation between ore grades is likely to deteriorate with distance. The problem is to know with some assurance the distance(s) over which this correlation exists and the precise nature of the variation in grade differences. Needless to say, none of these empirical methods can be accorded confidence levels so that there is no means of knowing how accurate the estimates are.

Kriging provides a system of weights which, when applied to co-related grade values, minimises the estimation variance of the point or block grade, as defined by:

$$\sigma\frac{2}{E} = E\left[\left(G_v - \bar{G}_{s_i}\right)^2\right]$$

Where $\sigma\frac{2}{E}$ = estimation variance
E = expected value
Gv = true (unknown) grade at point or block
$\bar{G}_{s_i}$ = Kriging estimator of the grade at point or block

Whereas some (rare) sets of assay values follow a normal distribution, many more have a lognormal or even more skewed distribution. In the latter case, the data is well skewed, as shown in the graph of the distribution of gold grades taken every 3 metres at a gold mine.

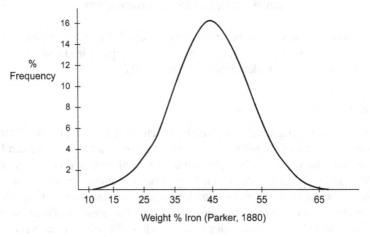

Weight % Iron (Parker, 1880)

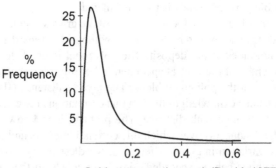

Gold concentration, oz/t (David, 1977)

However, the distribution of grade values, per se, tells us nothing of the independence or inter-dependence of ore grades within a deposit. If, for example, two neighbouring specimens are independent of one another then knowing the grade at one of them will tell us nothing of the grade to expect at the other. In some gold deposits, the grade at one point may tell us little of what to expect 5 metres or so away. There are, thankfully, many other deposits that demonstrate strong interdependence of grades over some distance as defined by the classical semivariogram.

The variance of the distribution of grade differences at fixed distances apart in a mineral deposit as indicated by its semivariogram will be a measure of the spread of the error distribution, which, in turn, defines the confidence to be placed on grade estimates. In other words, a low estimation variance is indicative of an estimate close to the true value.

Solution of the general problem (ordinary kriging)

Let us consider a block of ore of volume, V, having a true (but unknown) grade of G_v and a series of n specimens within its sphere of influence with recorded grades of G_{s_i} (I = 1,2, $\ddots$ . . n). The challenge is to find the set of weights, λ_i(I = 1,2, . . . n) which will make the weighted average :

$$\bar{G}_{s_i} = \frac{\sum \lambda_i G_{s_i}}{\sum \lambda_i}$$

the best estimator.

How can this be done? Well, the usual means of assessing the quality of an estimation is by computing the variance of the error associated with the estimation, i.e. $\sigma\frac{2}{E}$, given by:

$$\sigma\frac{2}{E} = \text{VAR}\left(E_i\right) = \text{VAR}\left[G_v - \bar{G}_{s_i}\right]$$

$$= \text{VAR}\left[G_v\right] - 2\text{COV}\left[G_v, -\bar{G}_{s_i}\right] + \text{VAR}\left[\bar{G}_{s_i}\right]$$

$$= \sigma\frac{2}{v} - 2\sum_i \lambda_i \sigma_{vs_i} + \sum_i \sum_j \lambda_i \lambda_j \sigma_{s_i s_j}$$

Where $\sigma\frac{2}{v}$ = VAR [Gv], i.e. the variance of the grade of block V.

σ_{vs_i} = COV [$Gv, \bar{G}_{s_i}$], i.e. the covariance of the grade of block V and the grade of specimen S_i

$\sigma_{s_i s_j}$ = COV of the grades of specimens S_i and S_j.

what we now need to do is to devise a means of minimising the variance of the estimation variance, σ_E^2, and by doing so ascertain the best values for the weights to be applied to the specimen values.

This can be done by differentiating the function, Q, of λ_i's and equating it to zero. However, when there is constraint, such that $C = o$, the Lagrange principle states that the relationship

$$F = Q + 2\,\mu C$$

should be minimised, where μ is a new unknown, the so-called Lagrange multiplier. In this case the relevant equation is therefore:

$$F = \sigma\frac{2}{E} - 2\mu\left(\sum_i \lambda_i - 1\right)$$

so that

$$F = \sigma\frac{2}{v} - 2\sum_i \lambda_i \sigma_{vs_i} + \sum_i \sum_j \lambda_i \lambda_j \sigma_{s_i s_j} + 2\mu\left(\sum_i \lambda_i^{-1}\right)$$

and the partial derivatives are:

$$\frac{\partial F}{\partial \lambda_i} = -2_{vs_i} + 2\sum \lambda_j \sigma_{s_i s_j} + 2\mu = 0$$

$$(I = 1,\ldots\ldots n)$$

$$\frac{\partial F}{\partial \mu} = \sum_i \lambda_i - 1 = 0$$

This procedure has therefore produced a linear system of $n + 1$ equation with $n + 1$ unknowns which are the unknown weights and μ. These equations can be written in their more general form:

$$\sum_j \lambda_j \sigma_{s_i s_j} + \mu = \sigma_{vs_i}$$

And

$$\sum_j \lambda_i = 1$$

Or in matrix form:

$$[\Sigma]\ [A]\ =[D]$$

Where Σ, A and D are, respectively (with σ_{ij} replacing $\sigma_{s_i s_j}$)

$$\Sigma = \begin{vmatrix} \sigma_{11} & \sigma_{12} & \cdots\cdots & \sigma_{1n} & 1 \\ \sigma_{21} & \sigma_{22} & \cdots\cdots & \sigma_{2n} & 1 \\ | & & & & \\ | & & & & \\ | & & & & \\ | & & & & \\ | & & & & \\ \sigma_{n1} & \sigma_{nz} & \cdots\cdots & \sigma_{nn} & 1 \\ 1 & 1 & \cdots\cdots & & 0 \end{vmatrix}$$

$$A = \begin{vmatrix} \lambda_1 \\ \lambda_2 \\ | \\ | \\ | \\ \lambda_n \\ \mu \end{vmatrix}$$

$$D = \begin{vmatrix} \sigma_{vs_1} \\ \sigma_{vs_2} \\ | \\ | \\ | \\ | \\ | \\ \sigma_{vs_n} \\ 1 \end{vmatrix}$$

The solution in terms of the unknown weights is therefore:

$$[A] = [\Sigma]^1 [D]$$

The problem is soluble since all the σ's (i.e. all variances and covariances) are obtainable from the semivariogram.

Point kriging

In this case, the problem is to estimate the grade at a point rather than that of a block. The previous equations can therefore be simplified, since instead of having to take account of the covariance of the grade of a sample and the grade of a block σ_{vs_i}, we only have to consider covariances of specimens $\sigma_{s_o s_i} = \sigma_{oi}$, assuming S_o is the point to be estimated, thus:

$$
\begin{vmatrix} \lambda_1 \\ \lambda_2 \\ \\ \\ \lambda_n \\ \mu \end{vmatrix}
=
\begin{vmatrix}
\sigma_{11} & \sigma_{12} & \cdots & \sigma_{1n} & 1 \\
\sigma_{21} & \sigma_{22} & \cdots & \sigma_{2n} & 1 \\
\\
\sigma_{n1} & \sigma_{nz} & \cdots & \sigma_{nn} & 1 \\
1 & 1 & \cdots & 1 & 0
\end{vmatrix}^{-1}
\cdot
\begin{vmatrix}
\sigma_{01} \\
\sigma_{02} \\
\\
\sigma_{0n} \\
1
\end{vmatrix}
$$

If there are only two specimens to be considered having grade values of G_1 and G_2 we would therefore end up with the following three equations:

$$\gamma(G_1 G_1)\lambda_1 + \gamma(G_1 G_2)\lambda_2 + \mu = \gamma(G_1 V)$$
$$\gamma(G_2 G_1)\lambda_1 + \gamma(G_2 G_2)\lambda_2 + \mu = \gamma(G_2 V),$$
$$\lambda_1 + \lambda_2 \qquad\qquad = 1$$

Examples of point kriging

Following are two very simple examples of point ordinary kriging. They are a little unrealistic since only two point grade values are used in the estimation of the grade at another point. This has been done in order to demonstrate the method of kriging without the complication of having too many simultaneous equations.

Example 1

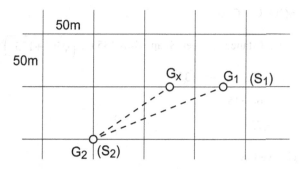

The isotropic orebody demonstrates a spherical semivariogram with the following characteristics:

Range of influence = a = 90m
Spatial variance = C = 60
Nugget effect = C_o = 15

Given the ore grades for drill-core specimens S_1 and S_2 are G_1 = 3.6 g/t and G_2 = 4.1 g/t, calculate the most probable value of ore grade at point x.

The method of kriging defines the following linear system of equations for this situation:

$$\lambda_1 \gamma(G_1 G_2) + \lambda_2 \gamma(G_1 G_2) + \mu = \gamma(G_x G_1)$$
$$\lambda_1 \gamma(G_2 G_1) + \lambda_2 \gamma(G_2 G_2) + \mu = \gamma(G_x G_2)$$
$$\lambda_1 + \lambda_2 = 1$$

since the variance of a specimen value with itself is zero (i.e. I = o) we have that

$$\gamma(G_1 G_1) = \gamma(G_2 G_2) = 0$$

The covariance between the two specimen values (i.e. G1 and G2) and between each of these and the point to be estimated can be obtained from the spherical model, i.e.:

$$\gamma(l) = C\left[\frac{3}{2} \cdot \frac{l}{a} - \frac{1}{2} \frac{l^3}{a^3}\right] + C_0$$

For $I \leq a$

And $\gamma(l) = C + C_0$ for $I > a$

Thus since the distance between S_1 and S_2 is 135 m, $\left(\sqrt{50^2 + 125^2}\right)$ we have

$$\gamma(G_1 G_2) = \gamma(G_2 G_1) = \gamma(135)$$
$$= 60 + 15$$
$$= 75$$

For $\gamma(G_X G_1)$ we have

$$\gamma(G_X G_1) = \gamma(50$$
$$= 60\left[\frac{3}{2} \cdot \frac{50}{90} - \frac{1}{2} \frac{50^3}{90^3}\right] + 15$$
$$= 60 \times .744 + 15$$
$$= 59.6$$

For $\gamma(G_1 \times S_2)$ the distance between the points is $\sqrt{75^2 + 50^2} = 90\,m$, so that:

$$\gamma(G_X G_2) = \gamma(90$$
$$= 60 + 15$$
$$= 75$$

Consequently the original equations are reduced to :

$$75\lambda_2 + \mu = 59.6$$
$$75\lambda_1 + \mu = 75$$
$$\text{and } \lambda_1 + \lambda_2 = 1$$

Solving these equations leads to:

$$\lambda_1 = 0.6$$
$$\lambda_2 = 0.4$$

Using these weighting coefficients the most probable value of the ore grade at point X is given by:

$$G_x = 0.6 \times 3.6 + 0.4 \times 4.1 = 3.8\,g/t$$

Using the inverse distance squared (ID^2) method gives:

$$G_x = \frac{\frac{1}{50^2}.G_1 + \frac{1}{90^2}.G^2}{\frac{1}{50^2} + \frac{1}{90^2}}$$

So that $\lambda_1 = 0.8$ and $\lambda_2 = 0.2$
and $G_x = 0.8 \times 3.6 + 0.2 \times 4.1$
$= 3.7\text{g/t}$

Example 2

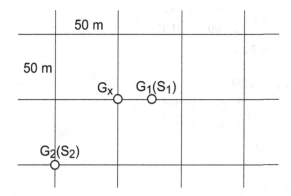

Given this isotropic orebody has a spherical semivariogram with the following characteristics:

Range of influence $= a = 100\text{m}$
Spatial variance $= C = 60$
Nugget effect $= C_0 = 15$

and the values of ore grade recorded for the drill-core specimens S_1 and S_2 are 3.0 and 4.4 g/t respectively, calculate the most probable ore grade at point X.

Once again, the linear system of $n + 1$ equations obtained by the method of kriging are:

$$\lambda_1 \gamma(G_1 G_1) + \lambda_2 \gamma(G_1 G_2) + \mu = \gamma(G_x G_1)$$
$$\lambda_1 \gamma(G_2 G_1) + \lambda_2 \gamma(G_2 G_2) + \mu = \gamma(G_x G_2)$$
$$\lambda_1 + \lambda_2 = 1$$

In this case the distances between the various points are:

$$S_1 \text{ to } S_2 = \sqrt{75^2 + 50^2} = 90\,m$$
$$X \text{ to } S_1 = 25\,m$$
$$X \text{ to } S_2 = \sqrt{50^2 + 50^2} = 71\,m$$

Thus the covariance between the point grade values can be derived from the spherical model:

$$\gamma(G_1 G_2) = \gamma(G_2 G_1) = \gamma(90)$$
$$= 60 \left[\frac{3}{2} \cdot \frac{90}{100} - \frac{1}{2} \cdot \frac{90^3}{100^3} \right] + 15$$
$$= 74.4$$

$$\gamma(XG_1) = \gamma(25)$$
$$= 60 \left[\frac{3}{2} \cdot \frac{25}{100} - \frac{1}{2} \cdot \frac{25^3}{100^3} \right] + 15$$
$$= 37$$

$$\gamma(XG_2) = \gamma(71)$$
$$= 60 \left[\frac{3}{2} \cdot \frac{71}{100} - \frac{1}{2} \cdot \frac{71^3}{100^3} \right] + 15$$
$$= 67.8$$

Consequently, the original equations are reduced to:

$$74.4\lambda_2 + \mu = 37$$
$$74.4\lambda_1 + \mu = 67.8$$
$$\lambda_1 + \lambda_2 = 1$$

Using these weighting coefficients we have that the most probable value of the ore grade at X is given by:

$$G_x = 0.71 \times 3.0 + 0.29 \times 4.4$$
$$= 3.4 \text{ g/t}$$

Applying the inverse distance squared (ID²) method gives:

$$G_x = \frac{\frac{1}{25^2} \cdot G_1 + \frac{1}{71^2} \cdot G_2}{\frac{1}{25^2} + \frac{1}{71^2}}$$

So that $\lambda_1 = 0.9$ and $\lambda_2 = 0.1$

and $G_x = 3.1$ g/t

Example of block kriging

Brooker (1980) presented an interesting example of block kriging based on an isotropic horizontal tabular deposit, which had been sampled by vertical drillholes on a square grid pattern.

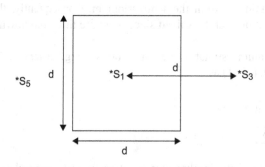

The semivariogram for the deposit is assumed to be of the spherical type with the following characteristics:

Range of influence = a = 2d
Nugget effect = C_o = 0
Sill value = $C + C_o$ = 1

In this case, because of the symmetry of the specimens and the isotropy of the deposit, the same weight must be apportioned to each of the grades recorded for specimens S_2, S_3, S_4 and S_5. It follows, therefore, that the best estimate of the grade for the block can be reduced to a problem involving only two specimen values, i.e.:

$$\bar{G}_k = \lambda_1 G_{s_1} + \lambda G_{s_x}$$

Where $G_{s_1} = S_1$
and $G_{s_x} = 0.25(S_2 + S_3 + S_4 + \dot{S}_5) = S_X$

It also follows that since there are only two weights to be considered and bearing in mind the non-bias condition, the kriging system of equations becomes:

$$\gamma(S_1 S_1)\lambda_1 + \gamma(S_1 S_x)\lambda_2 + \mu = \gamma(S_1 v)$$
$$\gamma(S_x S_1)\lambda_1 + \gamma(S_x S_x)\lambda_2 + \mu = \gamma(S_x v)$$
$$\lambda + \lambda_2 = 1$$

It should, of course, be remembered that the terms $\gamma(S_1 S_x)$, $\gamma(S_1 S_1)$, etc., are numbers obtainable from the semivariogram. Consequently, these simultaneous equations can be solved since there are three unknowns and three equations.

The minimum estimation variance, or 'kriging variance', is given by Journel et al. (1981) as:

$$\sigma_K^2 = E\left[\left(Gv - \bar{G}_K\right)^2\right]$$
$$= \sum_i \lambda i \gamma(Si, v) + \mu - \gamma(v, v)$$

Subtracting the first simultaneous equation from the second and using the non-bias condition leads to:

$$\lambda_1 = \frac{\gamma(S_x v) - \gamma(S_1 v) + \gamma(S_1 S_x) - \gamma(S_x S_x)}{2\gamma(S_1 S_x) - \gamma(S_1 S_1) - \gamma(S_x S_x)}$$

We now have to evelute the γ terms knowing that:

$$\lambda(h) = C\left[\frac{3h}{2a} - \frac{h^3}{2a^3}\right] \quad h < a$$

and $\lambda(h) = C \quad h > a$

It follows, therefore, that:

$$\gamma(S_1 S_x) = \gamma(S_x S_1) = \gamma(d$$

$$= \left[\frac{3d}{4d} - \frac{d^3}{16d^3}\right]$$

$$= 0.688$$

It can also be assumed that:

$$\gamma(S_1 S_1) = 0$$

Whereas $\gamma(S_x S_x)$ is given by :

$$\gamma(S_x S_x), = 0.25\left[2\gamma\left(d\sqrt{2}\right) + \gamma(2d)\right]$$

and

$$2\gamma\left(d\sqrt{2}\right) = 2 \times 1\left[\frac{3d\sqrt{2}}{4d} + \frac{d^3\left(\sqrt{2}\right)^3}{16d^3}\right] = 1.766$$

$$\gamma(2d), = 1.0$$

So that:

$$\gamma(S_x S_x) = \frac{2.766}{4} = 0.692$$

As Brooker (1980) states, the terms $\gamma(S_1\ v)$ and $\gamma(S_x\ v)$ can be written in terms of a so-called auxiliary function which he defines as H(m, n).

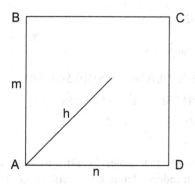

H(m, n) is the average value of $\gamma(h)$ when one end of the vector h is fixed at one corner of the block, A, and the other covers the area mn of the square

Table 5.1 H(m, n) values for spherical model, a = C = 1

n	0.05	0.15	0.25	0.50
0.05	0.057	0.123	0.193	0.364
0.15	0.123	0.171	0.231	0.389
0.25	0.192	0.231	0.282	0.425
0.50	0.364	0.389	0.425	0.535
0.75	0.513	0.531	0.558	0.644
1.00	0.628	0.642	0.662	0.728
1.50	0.752	0.761	0.775	0.819

ABCD. This can be tabulated for any spherical model and standardised tables are available that can be converted to the relevant sill and range of influence. For a spherical model all distances are relative to the range of influence, a. Hence the value of H(m, n) for a model with range, a, is the same as the value of $H(^m/_a, ^n/_a)$ for a model with range 1. Consequently, Table 5.1 with range 1 covers all spherical models.

It follows that since $\gamma(S_1\ v) = H(0.5d, 0.5d)$ where $d = ^a/_2$, the term required is H(0.25a, 0.25a) with a range of a, or H(0.25, 0.25) with a range of 1. From Table 5.1 we have that:

$$\gamma(S_1\ v) = H(0.25, 0.25) = 0.282\,C$$

And since C = 1

$$\gamma(S_1\ v) = 0.282.$$

Similarly, we have for $\gamma(S_x\ v)$:

$$\gamma(S_x v) = 1.5\,H(1.5d, 0.5d) - 0.5\,H(0.5d, 0.5d)$$
$$= 1.5\,H(.75a, .75a) - 0.5\,H(.25a, .25a)$$
$$= 1.5 \times 0.558 - 0.5 \times 0.282$$
$$= 0.696.$$

Brooker goes on to say that tables for F(m, n) – the average of γ(h) as both end of vector h independently cover ABCD, i.e. F(m, n) = γ(ABCD, ABCD), are also available. Using these tables the value of γ(v, v) = F(d, d) was found to be: 0.376.

Knowing all these constants it is a simple arithmetic task to determine the values of γ_1 and γ_2 which are:

$$\gamma_1 = 0.600$$
$$\gamma_2 = 0.400.$$

The minimum estimation or kriging variance is given by:

$$\sigma_K^2 = \sum \lambda_i \gamma(s_i v) + \mu - \gamma(v,v)$$

and

$$\mu = \gamma(S_1 v) - \lambda_1 \gamma(S_1 S_1) - \lambda_2 (S_1 S_x)$$
$$= 0.282 - 0 - 0.688 \times 0.4$$
$$= .007$$

So that

$$\sigma \frac{2}{K} = .078$$

It follows therefore, that the best estimate of the grade of the block is given by:

$$\bar{G}_k = 0.6 G_{s_1} + 0.1 \left(G_{s_2} + G_{s_3} + G_{s_4} + G_{s_5} \right)$$

Comparing this estimate with the result obtained by using the polygonal method when all the weight is applied to the central value, we have that the variance of the error is given by:

$$\sigma \frac{2}{E} = 2 \times 1 \times .282 - 0.376$$
$$= 0.188$$

Which is almost 2.5 times the kriging variance. In other words, the confidence limits for the polygonal method would be more than 50% wider than those for kriging.

Let us now consider the inverse distance (ID) methods and in doing so let's assume that the specimen S_1 is not at the centre of the block, but at a distance $\frac{d\sqrt{2}}{4}$ from the centre of the block (i.e. one quarter of the diagonal).

Hence, for the inverse distance method:

$$\lambda_1 = \frac{4}{d\sqrt{2}}$$

and $\lambda_2 = \frac{4}{d}$, since there are 4 specimens at a distance, d, from the centre of the block.

Therefore:

$$\frac{\lambda_1}{\lambda_2} = \frac{1}{\sqrt{2}} = .707$$

and since $\lambda_1 + \lambda_2 = 1$

$$\lambda_1 = 0.414$$

and $\qquad \lambda_2 = 0.586.$

The estimation error is given by:

$$\sigma\frac{2}{E} = 2\sum_i \lambda_i \sigma(S_i v) - \gamma(vv) - \sum_i \sum_j \lambda_i \lambda_j \gamma(s_i s_j)$$

Which gives:

$$\sigma\frac{2}{E} = 2\left[\lambda_1 \gamma(S_1 v) + \lambda_2 \gamma(S_x v)\right] - \gamma(v,v) - \left[\lambda_1 \lambda_2 \gamma(S_1 S_x) + \lambda_2 \gamma(S_x S_x)\right]$$
$$= 2 \times .525 - .376 - .573$$
$$= .101.$$

For the inverse distance squared (ID^2) method, we have that:

$$\lambda_1 = \frac{16}{2d^2} \text{ and } \lambda_2 = \frac{4}{d^2}$$

So that:

$$\frac{\lambda_1}{\lambda_2} = 2$$

making:

$$\lambda_1 = .667 \text{ and } \lambda_2 = .333.$$

Hence the estimation error is:

$$\sigma\frac{2}{E} = .081$$

Lastly, for the ID^3 method we have that:

$$\lambda_1 = \frac{64}{2\sqrt{2}.d^3} \text{ and } \lambda_2 = \frac{4}{d^3}$$

So that

$$\frac{\lambda_1}{\lambda_2} = 5.66$$

Making:

$$\lambda_1 = 0.85 \text{ and } \lambda_2 = 0.15$$

So that the estimation error is:

$$\sigma \frac{2}{E} = 0.144$$

Brooker continued to use this simplified model to investigate the effects of changes in the characteristics of the orebody on the estimation process. He, therefore, carried out similar calculations for different variograms. All the variograms were kept at the same sill value, i.e. $C + C_o = 1$, but the range of influence and relative nugget effect (C_o/C) were varied.

The three semivariograms chosen represent orebodies whose mineralisation goes from being evenly dispersed and continuous to where the mineralisation is highly erratic, as demonstrated in the following diagrams.

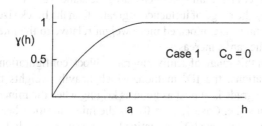

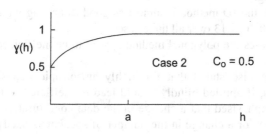

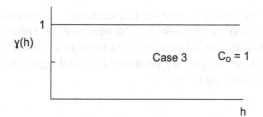

In the first two cases (i.e. $C_0/C = 0$ and 0.5) three ranges of influence were considered, i.e. $a = d$, $2d$ and $10d$, and finally an orebody with a pure nugget effect (i.e. $C_0 = 1$) was analysed. The result of Brooker's calculations are given in Table 5.2.

There are, as Brooker pointed out, a number of lessons to be learnt from this exercise, which can be summarised as follows:

- Kriging weights change in accordance with the semivariogram of the deposit, whereas the weights associated with the other methods are determined solely by the relative positions of the specimens and the block and remain constant.
- In kriging the weight allocated to the well-placed internal specimen is highest when the nugget effect is least. As the nugget effect increases, the mineralisation becomes more erratic until, in the extreme limit of a pure nugget effect, all specimens have the same weight.
- Providing the range of influence is greater than the block size, the external specimens are allocated more weight relative to the internal specimen as the range increases.
- Based on this range of semivariograms, block configuration and specimen locations, the ID^2 method, which heavily weights the internal specimen grade is almost as good as kriging when the mineralisation is continuous (i.e. Case 1, $C_0 = 0$). As the mineralisation becomes more erratic, however, the ID^2 estimation becomes increasingly less effective than kriging.
- Overall, the ID method is almost as good as kriging with R varying from 1.01 to 1.13 over all three cases.
- In all cases the polygonal method is seen to be inefficient relative to kriging.
- Brooker also stated that for highly anisotropic deposits, the ID method, if applied blindly, would lead to inefficient estimates. He also emphasised that a change in the data configuration relative to the block or a change in the number of specimens used in the estimate can change the efficiency of the estimating technique relative to kriging.
- It is also important to point out that the weights determined by any of the ID methods are dependent on the location of the specimens relative to the block and are independent of the variogram. Brooker also pointed out that these weights are determined without regard to the relative position of the specimens.

Table 5.2 Comparison of kriging, inverse distance and polygonal weighting schemes

Nugget Effect	Range	Kriging			Polygonal				Inverse Distance				Inverse Distance Squared			
		λ_1	λ_2	σ_E^2	λ_1	λ_2	σ_E^2	R	λ_1	λ_2	σ_E^2	R	λ_1	λ_1	σ_E^2	R
$C_o = 0$	a = d	.541	.459	.144	1	0	.407	1.68	.414	.586	.164	1.07	.667	.333	.164	1.07
	a = 2d	.600	.400	.079	1	0	.188	1.55	.414	.586	.102	1.14	.667	.333	.082	1.02
	a = 10d	.573	.427	.015	1	0	.037	1.57	.414	.584	.018	1.10	.667	.333	.016	1.03
$C_o = 0.5$	a × a	.370	.630	.208	1	0	.704	1.84	.414	.586	.211	1.01	.667	.333	.318	1.24
	a = 2d	.341	.659	.175	1	0	.594	1.84	.414	.586	.180	1.01	.667	.333	.277	1.26
	a = 10d	.233	.767	.115	1	0	.518	2.12	.414	.586	.138	1.09	.667	.333	.244	1.46
$C_o = 1$		0.2	0.8	0.2	1	0	1.0	2.24	.414	.586	.257	1.13	.667	.333	.472	1.54

Note: R = σ_E / σ_K is the ratio of the normal confidence interval of the particular technique to that for kriging.

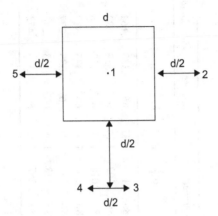

This he demonstrated using the configuration of specimens shown in the diagram. The ID method gives the same weights to specimen grades at 2 and 5 and slightly less weight to each of the specimen grades at locations 3 and 4. It is obvious that specimen grades at 3 and 4 relate to the same localised area and should be combined and be allocated the same weight as specimen grades at 2 and 5. The ability of kriging to take account of the fact that specimen values 3 and 4, because of their close proximity, should be assigned weights considerably less than those allocated to specimen values at 2 and 5, is demonstrated in the following values based on a spherical semivariogram with $C_o = 0$, $a = 2d$, and $C = 1$.

This simple example demonstrates kriging's ability to automatically take into account the fact that samples 3 and 4 are close to one another. Consequently, they are allocated weights considerably less than those allocated

Table 5.3 Comparison of ID and kriging weights

Point	Weight	
	ID method	*Kriging*
1	.418	.652
2	.148	.130
3	.143	.044
4	.143	.044
5	.148	.130
Estimation variance	.152	.095

to the specimen values at locations 2 and 5. The estimation variance also shows that by ignoring the spatial distribution of the specimens the confidence limit is increased by 60%.

Concluding comments

In many respects, the definition of mineral resources and ore reserves at the feasibility study stage is the most important element in a complex assessment of the mineral asset (i.e. the orebody). Not only is it necessary to estimate the quantity and quality of the extractable ore at this stage, but assumptions have also to be made regarding other features of the deposit, notably its metallurgical properties, geological structure and geomechanical environment.

It is at the feasibility study stage that Boards of Directors decide whether or not to develop a new mine. Once having decided to proceed with mine development, other important decisions regarding the location of shafts and/ or other entries, mining systems and treatment processes also have to be made.

All these decisions have unfortunately to be made on the basis of very limited data and it is not surprising, therefore, that many mines have experienced problems due to overestimation of ore grades, inadequate definition of geological faulting and/or geomechanical issues. Once mining commences, the amount of data available increases exponentially, although the new closely spaced data is always close to the current extraction areas. Grade control drilling is implemented immediately prior to mining. In open cut gold mines, these drillholes can be as close as 5m to each other.

Localised estimates of ore reserves are required for the 5-, 10- and 15-year cash flow projections, which are made at the feasibility study/planning stage, and if no more detailed information is available about these areas, scheduled for early exploitation, substantial errors could be associated with such forecasts.

A Preliminary Feasibility Study (PFS) is usually undertaken once sufficient indicated resources are established. A PFS will establish the time taken to return profits equal to the initial capital investments (payback period) amongst other things. This payback period can be applied as a useful guide to determine the extent of infill drilling programs. The ore scheduled to be mined during the payback period is drilled at a close enough hole spacing to obtain measured mineral resources (i.e. specimens are close enough to produce good variograms). Having well-defined (proven) ore reserves for the initial years of mining gives investors more confidence that they will at least have their capital returned.

As stated earlier, it is unfortunate that so many crucial decisions have to be made when the data available is so limited in both quantity and quality. The quantity of data is largely dependent on the number and proximity of specimens obtained primarily from drillholes, whereas the quality depends on the extent to which specimens are truly representative of the orebody and the care taken in their subsequent analysis. Of course drilling is very expensive, and there is no guarantee of a return on the expenditure – only if mining occurs one day, and that it is profitable.

Statistically there are some issues worthy of further comment. If the process of sampling an orebody is to be statistically meaningful, the samples should be randomly obtained and similar in size. (Refer to Chapter 3 for a full discussion of this topic.) Whereas the first issue has been adequately dealt with, the issue of sample size is also important, because as sample size increases the variance of the distribution of values decreases. Obviously, mining companies have to balance the need for quality and precision in defining prospective orebodies, which increase as sample size increases, with the need to obtain the best overall view of the deposit at an acceptable cost, which may be best achieved by a larger number of smaller-sized samples.

Nevertheless, once having obtained a good overview of the prospective orebody from surface drilling, it is desirable and sometimes essential to enhance this by sinking a test pit or shaft to obtain a large bulk specimen of ore for more detailed analysis and metallurgical test work.

In assessing drill core samples, the mineralised intersections are initially identified by visible evidence of mineralisation. Specimens may also be examined beyond these visible limits to explore for unsuspected mineral/metal values.

If an orebody is to be based on a 'cut-off-grade' (COG), assay values will be used to define the mineralisation above this limit in an attempt to define the economically mineable ore.

It is obvious, therefore, that ore reserve estimation is both a science and an art, since the experiential judgement of exploration geologists and ore reserve estimators are important elements in the overall process. A Perth-based geostatistician, Jacqui Coombes, published a useful book in 2008 titled *The Art and Science of Resource Estimation – A practical guide for geologists and engineers*, which develops many of the concepts discussed here in greater detail.

As stated earlier, one of the crucial decisions in the information gathering process is the spacing of drillholes, since a delicate balance exists between the need for more information and the cost of drilling.

Call (1979) claims that the relationship between the cost of drilling and drillhole spacing for drilling vertical holes into a flat-lying orebody is given by the following equation:

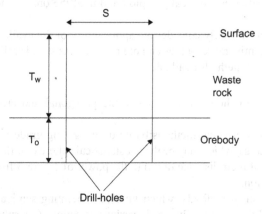

$$\text{Cost}/1{,}000\text{m}^3 = c/S^2\,(1 + T_w/T_o)\,1000$$
$$= x.c$$

Where c = drilling cost/metre
S = spacing of drillholes
T_w = thickness of waste rock
T_o = thickness of ore
X = drilling cost multiplier per 1,000m³ of ore

Tapp and Watkins (1990) stated that mineral prospecting is 'the world's biggest and best gambling business'. This is, in many respects, a truism, since the investment of funds in a mining venture is a decision based on imperfect knowledge. Even when substantial mineralisation is known to exist, the volume of ore and its grade variations can rarely be defined with a high degree of certainty.

In an attempt to overcome such difficulties, considerable emphasis is placed on the experience and knowledge of the estimators and evaluators. But, as Sorentino and Barnett (1994) state, one of the problems with this approach is the 'risk of logical failure, i.e. a tendency to see in information what one expects rather than what is really there'.

Another problem they highlight is the sheer complexity of mining projects, which is often reduced to manageable proportions by ignoring those issues considered by the evaluators to be least important.

Despite these difficulties in evaluating mining projects, the successes far outweigh the failures. Such successes are the best yardstick of the ability of exploration geologists, mining engineers and metallurgists to locate orebodies and thereafter economically exploit and treat the ore for the benefit of humanity.

Finally, let us briefly consider the applicability of the available methods of estimating mineral resources and ore reserves. As stated earlier, there are three groups of methods available:

- Classical methods, which include the polygonal and cross-sectional methods.
- Distance weighting methods based on weighting grade values within search areas, which may be the geostatistical sphere of influence, by the inverse of their distance raised to the power of 1, 2 or 3 from the point in question.
- Geostatistical methods, which involve generating semivariograms to define the range of influence in various directions (i.e. sphere of influence) before using geostatistical theory to determine the exclusive set of weighting coefficients to be applied to grade values within the range of influence. Geostatistical methods are many and varied including: simple kriging, ordinary kriging, indicator kriging, multiple indicator kriging, co-kriging, conditional simulation and uniform conditioning.

No matter what method is used, it is important to stress the significance of the value judgements that have to be made by the estimators. Such judgments are necessary due to the limited data and geological information available at the feasibility study stage and to a lesser extent at subsequent stages in the evaluation process. These judgements include the selection of the most appropriate method or methods of estimating mineral resources and/or ore reserves, defining the outer limits of the orebody and deciding on the recovery factor and level of dilution to be used in estimating ore reserves. Indeed, the paucity of data can be such, at the feasibility study stage, that meaningful semivariograms cannot be generated for some of the deeper or smaller domains of the orebody. If, for example, exploratory drillholes are too far apart (i.e. beyond the orebody's range of influence) the true nature of the ore grade distribution cannot be revealed and all that may be revealed is a pure nugget effect.

One of the prime decisions is, therefore, the selection of the most appropriate method to be used in estimating resources and reserves, and in making this decision. Carras (1990, 2001) suggests that:

- Orebodies with low coefficients of variation (CoV), say less than one, can be assessed with reasonable accuracy using the polygonal and/or the inverse distance cubed (ID^3) methods.
- Orebodies with mid-range CoVs, between 1 and 1.2, require smoothing of the data using either the inverse distance squared (ID^2) and/or geostatistical methods and hopefully the search area can be defined by the semivariogram(s).
- Orebodies with high CoVs are difficult to assess and, as Carras states, considerable emphasis must be placed on geological interpretation. In these instances, geostatistical methods may be difficult, if not impossible, to use and Carras suggests that classical methods with severe high-grade cutting can produce acceptable results.
- The assumption made in the polygonal method that the ore grade is constantly the central value within each polygonal block is inappropriate for most orebodies.
- Whereas classical methods are based on simple assumptions and can be easily applied and adapted to a wide range of orebodies, they have to be applied with care, especially when high-sample grade values have been recorded at some locations. Applying such high values over large areas can lead to serious over-estimation. Operating on the principle that it is better to underestimate ore reserves than create false high expectations, the high values are often reduced or cut to a more conservative value. Needless to say, such arbitrary decisions should be clearly explained and justified, otherwise the evaluation process becomes a questionable exercise.
- Whereas distance weighting methods are also easy to apply, the estimator has to decide which power (i.e. ID, ID^2 or ID^3) is the most appropriate to use. If the search area is not defined by geostatistical means, assumptions will also have to be made regarding its extent. Such arbitrary decisions demonstrate, once again, the important role of the estimator. A high power is appropriate for deposits with a low nugget effect, while a low power will provide better estimates where a high nugget is at play (e.g. gold deposits).
- Geostatistical methods are attractive because of their mathematical rigour. They involve the definition of semivariograms which are used to identify the most appropriate orebody model, e.g.

  - Models with a sill
  - Spherical, Exponential, Gaussian
  - Models without a sill
  - Linear, De Wijsian, Power Law

Those semivariograms with a sill can be used to define the search area. Thereafter, kriging can be used to determine the weighting coefficients to be applied to the grade values within the search area in order to estimate the most probable value of the grade at the central point.

The major problems with the use of geostatistics are the limited data often available at the exploration stage, which may not allow meaningful semivariograms to be generated, its complexity and its limited use for complex orebodies. For initial economic assessments, an inverse distance weighting method may be used in preference to the more complex and time-consuming process of kriging. However, once serious money is being spent on engineering studies, the cost of geostatistical analysis is small in comparison and a great value investment of funds.

Regardless of what method is used, there is no substitute for good-quality geological information and interpretations. In other words, resource and reserve estimations should always be made within the context of a well-informed interpretation of the orebody's geological structure and continuity. The geology is input to the grade model by wireframing, domaining, block modelling and validation using modern 3D mining software.

Goldsmith (2002) raises a number of important issues relating to the use made of resource and reserve estimates, such as their impact on financial reporting. They have considerable influence on mine life predictions and the calculations of depreciation and amortisation charges and consequently he emphasises the need for:

- Geologists and mining engineers being fully aware of the use others will make of their estimates. They should, therefore, give some indication of the confidence they place on the values assigned to the main variables, such as grade and dilution.
- Accountants understanding how, in a general sense, resource and reserve estimates are deduced, with particular reference to the assumptions made in the process.
- Company directors being aware of the in-house procedures for producing financial reports and the interaction between those who produce the resource and reserve estimates and those who subsequently use them. They should also have some understanding of the sensitivity of the financial results to the input variables.

In raising these issues, Goldsmith is implying that those responsible for providing resource and reserve estimates should be encouraged to demonstrate their level of confidence in the assumptions they are forced to make in arriving at such estimates.

The highest category of mineral resource in the JORC Code is the 'measured mineral resource', which is defined as 'that part of a mineral resource for which grade (or quality), densities, shape and physical characteristics are estimated with confidence sufficient to allow the application of 'modifying factors' to support detailed mine planning and final evaluation of the economic viability of the deposit'. It is based on detailed and reliable exploration, sampling and testing information gathered through appropriate techniques from locations such as outcrops, trenches, pits, workings and drillholes. The locations are spaced closely enough to confirm geological and/or grade continuity.

As far as ore reserves are concerned, the highest category in the JORC Code is a 'proved ore reserve', which is defined as the 'economically mineable part of a measured mineral resource'. It 'implies a high degree of confidence in the modifying factors'. The lower category is 'probable ore reserves' and these can either come up from indicated resources or be downgraded from measured resources – due to uncertainties in the 'modifying factors'. Ore reserves must include diluting materials and allowances for losses, which occur as the deposit is mined and is arrived at following appropriate assessments of all relevant factors. Ore reserves must be based on studies at the Pre-feasibility or Feasibility level, which demonstrate that the extraction could be reasonably justified. The JORC Code sets out the requirements for these studies in some detail. It also sets out all the modifying factors, which the Competent Person(s) (who takes legal responsibility) must address. The Competent Person must be an appropriately qualified and experienced mining professional and a member of the Australasian Institute of Mining and Metallurgy.

Inferred resources cannot become ore reserves.

These are 'deterministic' approaches to the quantification of resources and reserves, since it is based on a range of subjective judgements made by the competent person or persons.

In the mining industry the recoverable mineral reserves for a tabular deposit can be determined as follows:

$$\text{Recoverable reserves} = A \times T \times SG \times (1 + DF) \times RF_M \text{ tonnes}$$

where A = area of deposit
T = thickness of deposit
DF = dilution factor
SG = specific gravity of deposit
RF_M = mining recovery factor

If the extracted ore is to be treated in a mineral processing plant its recovery factor can be included to give the marketable recovered reserves:

$$\text{Marketable reserves} = A \times T \times SG \times (1 + DF) \times RF_M \times RF_P$$

where RF_P = mineral processing plant recovery factor

Each of these parameters could be defined as a distribution of values with its range being an indication of the estimator's confidence in the most likely value. This would in turn allow the probabilistic distribution for the ore reserves to be defined. One distinct advantage in using this approach is that it would force the estimators to define their confidence in the most likely values quoted for each parameter.

References and Relevant Bibliography

Appleyard, G.R. (2001). *An Overview & Outline*, Mineral Resource & Ore Reserve Estimation – AusIMM Guide to Good Practice, Monograph 23, The AusIMM, Melbourne.

Brooker, P.I. (1980). *Kriging, Geostatistics*, McGraw Hill, New York, 1980.

Call, D.C. (1979). Development Drilling. Chapter 3 in *Open Pit Mine Planning & Design* (Eds. J.T. Crawford, W.A. Husfrulia), American Institute of Mining, Metallurgical & Petroleum Engineers Inc., New York.

Carras, S. (1990). *Sampling Evaluation & Basic Principles of Ore Estimation*, Carras Mining & Associates, Perth.

Carras, S. (2001). *Let the Orebody Speak*, Mineral Resources & Ore Reserve Estimation – AusIMM Guide to Good Practice, Monograph 23, AusIMM, Melbourne..

Clark, I. (1980). *Practical Geostatistics,* Applied Science Publishers, London.

C:\Users\jbonnar\Desktop\EC assignments\15044-3093-Jones\15044-3093 Jones copyedit\15044-3093-For Copyedit\15044-3093-Ref Mismatch Report.docx - David, M. (1977). *Geostatistical Ore Reserve Estimation*, Elsevier Scientific Publishing Company, Amsterdam.

Various Authors (1980). *Geostatistics*, Compendium of Articles, McGraw Hill, 1980, New York.

Goldsmith, T. (2002). *Resources & Reserves: Their Impact on Financial Reporting, Valuations and the Expectations Gap*, CMMI Congress, Toronto, 2002.

Journel, A.G. and Huijbregts, C.J. (1981). *Mining Geostatistics*, Academic Press, London, 1978.

Lipton, I.T. (2001). *Measurement of Bulk Density for Resource Estimation, Mineral Resource & Ore Reserve Estimation: AusIMM Guide to Good Practice, Monograph 23.*

Open PitMine Planning & Design, (Eds. J.T. Crawford, W.A. Husfrulia), American Institute of Mining, Metallurgical & Petroleum Engineers Inc., New York, 1979.

Matheron, G. (1970). The theory of regionalised variables and their applications. Fascicule 5, Les Cahiers du Centre de Morphologie Mathematique, Ecole des Mines de Paris, Fontainebleau.

Rogers J.J.W and Adams A.S. 1963 Lognormality of Thorium Concentrations in the Conway Granite. Geochimica et Cosmochimica Acta, Vol 27, pp 775–783. Pergamon Press Ltd. Printed in Northern Ireland. (Also see Medelsohn F. November 1980. Some Aspects of Ore Reserve Estimation. University of Witwatersand, Economic Geology Research Unit, Information Circular No. 147, Sectin xii.

Royle, A.G. et al. (1980). *Geostatistics*, McGraw Hill, New York.

Rudenno, V. (1998). *The Mining Valuation Handbook*, Wrightbooks Pty Ltd.

Sorentino C. and Barnett D.W. 1994 Financial Risk and Probability Analysis in Mineral Valuation. Proceedings of VALMIN 94, the Australian Institute of Mining and Mettalurgy, Carlton, Australia.

SME MiningEngineering Handbook (Ed. I.A. Giren), Society of Mining Engineers of the American Institute of Mining, Metallurgical & Petroleum Engineers Inc., New York, 1973.

Tapp, B.A. and Watkins, J.R. (1990). *Energy & Mineral Resource Systems: An Introduction*, Cambridge University Press, New York & Melbourne.

Printed in the United States
by Baker & Taylor Publisher Services